Backyard Meat Production
How To Grow All The Meat You Need In Your Own Backyard
by Anita Evangelista

Loompanics Unlimited
Port Townsend, Washington

This book is sold for informational purposes only. Neither the author nor the publisher will be held accountable for the use or misuse of the information contained in this book.

Backyard Meat Production
How To Grow All The Meat You Need In Your Own Backyard
© 1997 by Anita Evangelista

All rights reserved. No part of this book may be reproduced or stored in any form whatsoever without the prior written consent of the publisher. Reviews may quote brief passages without the written consent of the publisher as long as proper credit is given.

Published by:
Loompanics Unlimited
PO Box 1197
Port Townsend, WA 98368
Loompanics Unlimited is a division of Loompanics Enterprises, Inc.
1-360-385-2230

Cover by Jim Siergey
Photos courtesy of Anita Evangelista and Tyny Goat Ranch,
   Oklahoma City, OK
Illustrations by Jim Blanchard

ISBN 1-55950-168-5
Library of Congress Card Catalog 97-74375

Contents

Chapter One
Why Grow Meat at Home?... 1

Chapter Two
The "Next White Meat" — Rabbit 9

Chapter Three
The Portable Egg Factory — Chicken......................... 35

Chapter Four
The "Wild Things" —
Quail, Pheasant, Guineas, Ducks............................... 57

Chapter Five
Porky's Pals — Mini Pigs 69

Chapter Six
The Milk & Meat Factory — Mini Goats 83

Chapter Seven
Leather, Feathers, Bones and Fibers......................... 103

Appendix
Butchering, Generic Style 117

Resources 125

Chapter One
Why Grow Meat at Home?

Once upon a time, this question would never have been asked. In agricultural parts of the world today, nobody asks this question, either... the poorer the nation, the less doubt the citizens have about growing backyard meat. It is only those of us who live in a prosperity and a luxury unmatched in history, who wonder: *why should I grow meat?*

For my family, the first part of the answer was financial — it is plainly cheaper to grow meat in your backyard than it is to buy it at the supermarket. Period. For example, we raised rabbits that cost us 38¢ per pound to put on the table... at the time the supermarket "fryers" were selling for $2.95 *per pound!*

Another part of the answer has to do with a more subtle issue: flavor. Backyard meat tastes BETTER! The meat has a more luscious texture, the natural juices are contained and percolate through the meat as it cooks, and the meat can be eaten as fresh as you like — or as aged as you like. You can raise animals that are super-lean or super-marbled and rich.

Backyard Meat Production

2

You can produce the best-tasting meat because you have complete control over every aspect of its production.

Other people have decided to become backyard meat-gardeners because of horror stories in the media: "E. Coli kills six who ate at fast-food restaurant," "Salmonella strikes again," "Bacteria in hamburger sickens family," "Brain disorder spread in beef," "New strain of trichina threatens pork consumers," "Resistant bacteria turns consumers away from meat-counters".... The news goes on and on, one frightening incident after another. What's a sensible person to do? Subsist on beans and rice??? Or make sure your meat is the most wholesome it can be by knowing where it came from and how it was handled?

When you grow meat, you "own" the product from birth to freezer. You know how the animal lived, how it was butchered, and how it was stored. If you handle the meat with simple hygiene, your chops and roasts will be as free from pathogenic bacteria as they can be. If you store it according to easy-to-follow directions, it will keep in prime condition until you are ready to use it. You will NEVER wonder, for even a second, if some careless handler has sneezed on your roast — or dropped it on a contaminated floor — or forgot to turn the temperature down far enough on the storage building. *You will be confident about your food!*

And then there is the issue of availability. While we do live in a nation that is prosperous by historical standards, we also have a country that is extremely vulnerable to sudden changes to the regular order. I'm thinking here of the severe and unusual weather that has plagued us for almost a decade — ice storms block transportation routes, floods cut off highways, earthquakes wreck supermarkets, the electricity is disrupted

Chapter One
Why Grow Meat at Home?

and goes out for weeks at a stretch, hurricanes frighten people into panic buying and shelves are emptied in a few hours... *and guess what disappears from markets first?* Bread, milk, and *meat*.

With a backyard meat garden, you won't have to make that worried rush to the supermarket, even if the skies darken and newspeople get frantic.

On the Other Hand...

I don't want to mislead you, though. Raising meat animals is a commitment that takes time, an investment of some start-up capital (which can be less than the cost of a family dinner at a fast-food restaurant, if you are good at "making-do"), and you will probably need to acquire new skills in the process. Realistically, you'll probably make some mistakes and will lose a few animals to preventable problems — but you'll learn quickly as the result.

You will also become friends with some of the animals who are destined for your table, and undergo the bitter-sweet experience of every generation of humans since animals were domesticated. You will learn important things about the mystery of life, and the mystery of death. After you have raised and butchered your own meat, you'll never again view those plastic-wrapped packages in the supermarket the same way. There are times when you will wonder about your own sanity, and try to remember WHY you ever decided to grow meat.

These are the inevitable stages of meat-gardening. After a few years, your entire perspective will have changed in a very interesting way....

Backyard Meat Production

4

The Vegetarian Issue

There's a pretty good chance that readers of this book will be committed carnivores or omnivores — eaters of meat. But in today's world, you will encounter more than one vegetarian who looks at your backyard meat garden as a pretty bizarre exercise.

Let me share a little story about our own "vegetarian" phase (a stage I'm happy to report we grew out of). Around 1978, with all the media attention (even then!) about the "terrible living conditions" of meat animals, we decided to give up meat. Like many of today's vegetarians, we felt that the "cruelty" of livestock production didn't justify our "meat-eating habit." Plus, we couldn't afford to buy the stuff.

So, we cut out almost all meat, except for eggs. After a few months, we must have been nutritionally starved for meat — because every time there was a commercial on TV that featured meats, we'd find ourselves salivating and remembering all those steak dinners and burgers that once were part of our lives. We just couldn't do it... we were vegetarian "failures"... we HAD TO HAVE OUR MEAT!!!

So, we made a compromise. If animals were being cruelly raised in "factory farms," we'd raise our own animals in model "cruelty-free" facilities in our teeny-tiny urban backyard.

We started with rabbits and chickens, progressed to quail, and eventually owned and raised all the types of critters covered in this book. That first butchering day, when our lovingly hand-raised rabbit met the Big Stick, still lives in my memory as a horror of botched technique and disgusting smells — and we left that bunny in the freezer for over a

Chapter One
Why Grow Meat at Home?

month before we could look at it as food and not a murder victim. But you know what? It tasted pretty darn good!

Over time, we got better and more efficient at the growing and at the butchering. We learned pretty quickly not to make "pets" out of the ones we were going to eat — by naming them things like "Sausage," and "Chops," and "Fried." All the animals lived comfortable lives free from stress and with plenty to eat. Their butchering was quick and as painless as we could make it. We found a method for tanning skins, and became fairly adept at turning rabbit skins into warm and elegant winter vests and booties for the children. We used every part of the animal, and the parts we couldn't use, we converted to dog food or fertilizer. Boy, oh boy, were we proud of our "cruelty-free" humane meat garden.

Until, that is, a vegetarian acquaintance visited and asked us that one pointed question that vegetarians ask: "You don't raise these cute bunnies to *eat* them, do you?"

The lesson we learned was that to the vegetarian world, it doesn't matter if you take good care of your animals — even if you treat the animals like members of your family — if you actually use them for *food!* Since then, we've come to believe that the relationship between people and animals is one that we can't escape, or change... animals are food for people, whether some people like it or not.

Bottom line: raise your meat well, enjoy your critters, and *revel* in your delicious meat! As long as you know you're doing it right, you don't have to explain it to anybody.

In the Zone

Your choice of animals in your meat garden will depend on two basic factors: what you enjoy, and where you live.

You've probably already got a good idea of what tastes good to you — rabbits, chicken, eggs, quail, wild fowl, duck, pork or even exotic chevon (goat, which is rather like beef). In this book, we'll examine each of these meat-garden regulars in detail.

But where you live may have a bearing on what you can raise. If you live within city limits, you may be subject to zoning laws which restrict "farm animals." Typically, this may refer to chickens, pigs, and cattle. However, in many cities this issue has been challenged by individuals who keep critters such as miniature pot-bellied pigs and miniature horses as pets. The best way to find out your city's zoning requirements is to call the Department of Animal Regulation (better known as The Dog Catcher), and ask them if you can keep the animal you want in your "residential" or "commercially" zoned area.

In some cities, suburban "horse" neighborhoods — where half-acre and quarter-acre properties are the norm — have few restrictions on meat gardens.

If you decide to keep chickens, you'll spare nearby neighbors the joys of early-morning "alarm clocks" by removing young roosters to the freezer. Rabbits are quiet, and we kept over 60 rabbits on a 40-foot by 70-foot urban property (which included a house and garden!) without the neighbors ever knowing we had them. Most of the animals in this book can be cage-raised in a garage or basement, without raising the neighbors' ire — or their suspicions!

Chapter One
Why Grow Meat at Home?

How to Plan Your Garden

Unless you have prior experience raising livestock (meat animals, not pet animals!), I suggest you begin with rabbits or chickens. Rabbits are excellent startup meat-garden elements because they are pet-like, cuddly, clean, and easy to care for. Chickens are good startups because they are small to begin with, and you can adjust their housing gradually until they are full-sized — plus, if you have trouble getting started on that first butchering, you still get the eggs! Both are inexpensive to buy, as you'll see in later chapters.

Game birds and ducks will be easy to raise after you've gained some experience with chickens — their needs are quite similar. But for keeping pigs or goats, should you decide to go that far, the experience you've had with rabbits will give you the skills you'll need with the bigger animals. Butchering a rabbit, for instance, is quite a bit similar to slaughtering a pygmy goat — and once you've learned the pattern on a rabbit, it's easier to apply it to larger meat animals.

Consider your property size, when you plan on acquiring animals of any kind. Remember that you will probably still want a lawn, flower beds, perhaps a vegetable garden, and a place to relax and for the kids to play. Situate your animals so that they are out of the direct sunlight (especially in hot regions of the country), and where they can get plenty of fresh air. Make certain that the animals aren't placed in low spots that fill with water during rainy spells, too!

Give some consideration to the possible uses you'll have for the animal's "by-products" — the hides, bones, guts, and fertilizer. The final chapter of this book offers some ideas and tanning methods, but your own situation can dictate new possibilities.

A Word of Caution

Finally, remember that your friends and neighbors may also be interested in fresh, home-gardened meats. There's nothing wrong with sharing, trading, or with selling them a live animal or two. But, you should not offer to sell a *butchered* carcass.

Here's the problem: if you sell a butchered carcass, the buyer cannot be sure if the animal was healthy or sick when it was processed — and you can be blamed if there is any human illness caused by the meat. But, if you sell a live animal, the buyer can determine by inspection that the animal is healthy-appearing and lacking anything that might cause human illness. *Then*, if you wish, you can butcher the animal for its *new owner*.

Yes, yes, yes — I know it's a technicality... but in today's litigious society, a little sensible caution goes a long way towards preventing hard feelings.

Chapter Two
The "Next White Meat" — Rabbit

The average American eats less than three pounds of rabbit per year. The average American is missing out on one of the finest, leanest, most easily digested, most easily raised, meats on this planet.

A typical female rabbit from meat-producing breeds (such as the New Zealand White, New Zealand Red, or the Californian) can readily and successfully produce 160 *pounds* of offspring every year. Three females and their male companion — a small collection of "pets" — can easily give you almost 480 pounds of rabbits yearly. (That's the same amount of meat produced by a cow, when she gives her one calf per year!) This works out to about 96 whole rabbits, or about two per week. If you really enjoy this meat, it's simple enough to double your "herd" to six females and one male, and double your production to four dinners per week.

Rabbit is a white skinless meat, averaging about 30% pure digestible protein (chicken is about 32%, sirloin of beef about 32%). It is considered very lean, less than 10% fat, and is so mild in flavor that it is similar to chicken breast in both texture and taste. Some hospital dietitians consider rabbit an ideal meat for convalescents, since it is so easy to digest. Rabbit can be substituted in virtually any recipe where chicken is used — and most people won't notice the difference.

A lovely pen of fryers "on the hoof."

Overview

Your rabbit-meat garden is most efficiently raised in cages, with each adult rabbit having its own cage. Rabbits, surprisingly, are fairly territorial and prefer having a "private space" to call their own. Some people who specialize in rabbit gar-

Chapter Two
The "Next White Meat" — Rabbit

dens keep their herd in a "community pen" — this is simply a wire-mesh lined enclosure at least eight feet by eight feet. Hay or other thick mulching material is piled up to two feet deep in the pen. One male and several females are penned together, with the male removed after two weeks. The girls will dig little burrows and have their young there.

Females, called "does," are bred once or twice to a male, a "buck," then 31-32 days later the babies will be born.

Rabbits need a "nest box" or special enclosure to have their babies. The birth, or "kindling," takes place inside the box, after the doe has pulled out a big fluff of her chest and neck hair to line the box. Mama rabbit will then spend most of her time outside of the nest, hopping in only every few hours for a couple of minutes to let the babies nurse. After two to three weeks, little miniature bunnies will be scrambling in and out of the nest on their own. At five to six weeks, the young ones can be put into a new cage of their own, and the mother rabbit rebred for her next litter. At six to eight weeks, the litter is transferred to the freezer — although some people let them get a little bigger and hold them until they are 12 weeks of age before butchering. When a rabbit of a meat-producing breed is eight weeks old, it will weigh about four and one-half to five pounds, and "dress out," or result after processing, at about two and one half to three pounds of fine, white meat.

Selecting the Meat Rabbit

Any old pet store rabbit can reproduce and provide you with something to eat — BUT! if you want efficient and cost-effective rabbit meat, you'll do better sticking to two or three

Backyard Meat Production

breeds which have been bred specifically for their meat. These breeds are the New Zealand White, an all-white rabbit with pink eyes (technically, an albino animal); the New Zealand Red, which is a reddish-brown color; and the Californian, a white-bodied rabbit with black ears, feet, tail, and nose, and pink eyes.

Rabbits of these breeds weigh between eight and one-half and 12 pounds at two years, and have rather fine bones and heavy muscling — this produces the most meat for your investment.

Of course, there's also a couple hundred other breeds of rabbits, including little three-pound miniature breeds such as the Netherland Dwarf (which tend to be pretty bad-tempered and irascible), and the huge 25-pound breeds like the Flemish Giant (which are slow-growing and hearty eaters). There's some archeological evidence that ancient Romans raised a type of rabbit that reached 45 pounds! There are breeds that have extremely fancy double-soft fur, like the Rex rabbits; and the "spinner's friend" breed called the Angora which produces a cozy fiber much in demand for knitting.

All these varieties of rabbit have their unique qualities — but if you want to eat them, stick to the good meat-types. In my experience, Dwarves are hard to breed and produce small litters, Giants eat too much compared to what they produce, and the fancy-hair varieties are far too thin to make a good meal.

A healthy rabbit (the only kind you should even consider) should have bright, clear eyes with no matter in the corners. The nose should be clean and not have any "snuffles" or snotty appearance. The coat should be smooth, glossy, thick

Chapter Two
The "Next White Meat" — Rabbit

and even, and shouldn't show areas of yellow staining. The insides of the rabbit's ears should be clean and be free of yellow crusty patches. There should be no sores on the rabbit's feet or anywhere else. Rabbits which make successful breeders are often the outgoing ones — they'll come up to you and sniff — and appear to be calm and inquisitive.

Our first rabbit came from a pet store. He looked a little scruffy, and had a slight head cold, but seemed okay to our unskilled eyes. He cost $12 as an eight-week old. By the time he was six months old, breeding age, he was showing signs of a significant genetic fault — maloccluded teeth that had grown out of his mouth so badly that he couldn't eat. Being new to rabbit care, we took him to a veterinarian, who was going to use a pair of wire-cutters to trim the teeth back to a normal length... but the shock of handling and transportation resulted in the poor rabbit's demise. This animal would NOT have been good breeding stock!

Our next rabbits were bought from an ad in a "recycler"-type paper, that read, "rabbitry sale." The rabbitry turned out to be a backyard meat garden that had become knee-deep in extra critters! We got seven healthy ten-pound New Zealand crosses, that were four months old... they cost $5 each. A month later, they were bred and went on to outproduce the averages (usually, the does had ten to 12 babies each litter!). Not one of them suffered from malocclusion, and several lived to ripe old ages (for a rabbit, that is) of six and seven years. They gave us the equivalent of two calves per year during that time.

The moral of this story is: stick with healthy commercial breeding stock of meat breeds!

The Garden Hutch

Housing for rabbits doesn't need to be new or fancy. But it is important that it be "critter proof" — strong enough to repel marauding neighborhood dogs, prowling cats, opossums, raccoons, snakes and rats, if these are problems in your area. For this reason, most home-made hutches are constructed from wire mesh (½" x ½") stapled or tacked to a 2´ x 4´ frame. Chicken wire may be adequate for the sides of the cage, if predators aren't a problem, but the floor of the cage should be that ½" mesh to allow good air flow and let the animals' waste fall through.

Double-decker cages, two above and two below, tucked into a niche in the garden.

Chapter Two
The "Next White Meat" — Rabbit

The photograph shows a double-decker rabbit hutch, consisting of four large cages. With a little ingenuity, this can be enlarged to a six-cage double-decker that will take up only about three feet by nine feet of floor space. An individual cage should be about 24" x 30", though a larger cage of 26" x 36" is often appreciated by the rabbits — especially when a mama rabbit is kept with eight to ten active little rabbits!

All-wire cages are perfectly fine for rabbits, as long as they are kept where they have shade. Rabbits must have shelter from direct sunlight, no matter where the cage is kept — with their thick coats and no shade, rabbits can expire very quickly from the heat of direct sunlight. However, they do need access to *some* sunlight — perhaps early morning sun, or in the late afternoon — as long as they can get out of the sun when they want to.

If you make your rabbit hutches, try to be certain that any wooden areas are covered by wire — bunnies will vigorously chew exposed wood, and can actually gnaw their way through a two-by-four! Also, rabbits will select one corner of their cage to use as the "toilet," and continual use may cause some wooden supports to get pretty soggy. Don't bother painting wooden hutches on areas that the rabbits can reach — they'll chew paint, too.

Rabbits are generally more sensitive to hot, humid weather than they are to cold weather. During the summer, your rabbits may stretch out and pant — make sure they have plenty of water and shade; take extra nest hair out of the nestbox to prevent the babies from becoming overheated. During the winter months, block the coldest airflows (we've used flattened cardboard boxes retrieved from supermarket throwaways), even down around the supports of the cage, to prevent freezing of bunnies and adult rabbit's ear tips.

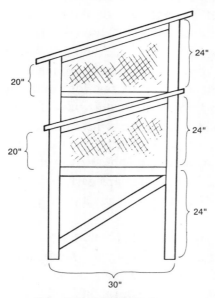

Side view, double-decker cage.

Used cages are just as good as new ones, with one caveat: the used cage must be cleanable. Make a solution of one part bleach to four parts water and spray or swab the cage thoroughly. Allow to air-dry fully. Swab again, and rinse with clear water. This treatment should destroy any disease organisms that may have been left behind by the previous occupant.

Generally speaking, rabbits don't try to break out of their cages — in fact, a calm rabbit might be content to stay in a cage that has no door on it, as long as food and water are adequate. Simple door latches, such as metal hook-and-eye closures, or even just a piece of curved wire attaching the door to the cage proper may be sufficient to keep rabbits in. Of course, a raccoon or possum won't be deterred by such an easy latch either! Little bunnies will find nooks and crannies to squeeze out through, so be certain even the smallest holes are patched with wire mesh.

Nest boxes can be homemade or store-bought. Typically, a nest box is two feet long by one foot high and one foot wide. One end is cut down to make it easy for the rabbit to get in and out, as shown in the illustration.

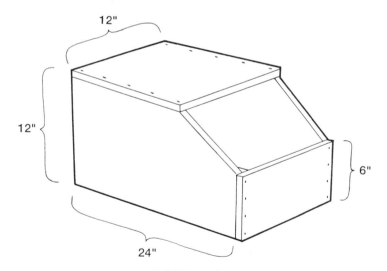

Rabbit nest box.

There are wood or metal nest boxes. Metal ones are easier to clean, but very cold during winter. Wood makes a good nest box because of its insulating properties during winter, but when the babies graduate out of the box, it is hard to scrape and clean out the "potty" the little ones leave behind. Use that one-part-bleach to four-parts water to slosh around the box, but be sure to rinse thoroughly and air-dry completely before putting it back to use — there should be no chlorine odor to it at all, which might damage baby-rabbit lungs. If you have a propane torch, you can scorch the inside of nest boxes as an alternative to bleaching them.

Nest boxes are put into the doe's cage three to four days before she is due to kindle. Any sooner, and she'll treat the nest box like a private toilet.

Feeds and Feeding

The agricultural industry has been thoughtful enough to spend a few decades researching and refining the types of feeds that help rabbits live well — the result is a pellet-shaped green feed, made principally from alfalfa hay, which contains additional vitamins, minerals, salts, and nutrients. It is the "perfect" rabbit food, and rabbits really seem to enjoy it.

The least expensive way to buy rabbit feed is in 50-pound bags. In the Midwest, 50 pounds of rabbit pellets runs about $5. In Los Angeles, the same bag might cost $12. If you have the time and transportation, a drive out to the country once a month to a "feed store" to pick up a couple bags may save you some money. (Check the telephone directory for outlying areas under "Livestock Feed.")

This pelletized feed comes as either 15% protein or 18% protein. Usually, for rabbits that are bred three or four times yearly, and for growing babies, the 15% feed is fine. The higher-protein 18% would be better for animals that are bred up to six times per year (it CAN be done!), or for show-rabbits that need to be in peak condition. One slight drawback of the higher- protein feed, in my experience, is that it can cause the rabbits to suffer from bowel irritations — from insufficient fiber, I think. If you decide to use 18%, be sure your rabbits get extra things to gnaw on (trimmings from apple trees, carrots, etc.).

Rabbit feeder, side view.

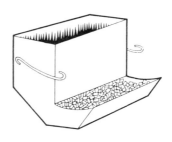

Rabbit feeder, ¾ view.

An adult buck or doe can be expected to eat between four and six ounces of feed daily. After giving birth, a doe should get as much as she will eat (which is rarely more than eight ounces to start with). Keep her feeder full when the bunnies start eating, because they will really go through the feed! The bunnies will grow best and fastest if they have constant access to food.

You can acquire special rabbit-feeders ($4), which are usually an "L"-shaped metal holder with screen on the bottom. This is clipped to the outside of the cage and the lower edge passes through to the inside of the cage; you fill the feeder from the outside, and the pellets fall past the screen (so any pellet dust is screened away), and into the interior of the cage for ready access to the rabbits. These are commercial-rabbitry feeders, which are both easy to use and help keep the feed clean and dry.

You can also prepare a homemade feeder from a metal coffee can, cut in half so that it is no more than three inches high. Smooth any rough edges, or bend them down. Wire this tightly to the inside of the cage (rabbits enjoy flipping these over). Every so often, clean out the buildup of green dust in the bottom of the feeder.

There's also a heavy ceramic crock ($4-$6) offered at some pet supply outlets which can be used as a feeder — but, for some reason, baby bunnies consider these an ideal spot to relieve themselves. If you wish to use these crocks, limit them to your bucks and unbred does!

Water is, perhaps, the most important nutrient your rabbits require — a day without water can be a serious compromise to the health of adults, and can result in lactating does "drying off" unexpectedly. The wisest approach to water is always to have fresh, clean water available to the rabbits.

Ceramic crocks ($4) make excellent waterers, and bunnies won't use them as a potty when crocks are filled with water. Any container that can be cleaned makes a suitable waterer, though — lightweight holders may have to be wired to cages to prevent overturning. Waterers made from plastic will be chewed on, too. For some reason no one has quite explained, rabbits will take certain items (such as tree trimmings, and leaves) and put them into their water — maybe, making tea? This seems to be a harmless habit, but if the water gets really yucky, the stuff should be removed. On hot days, some rabbits will sit next to their waterer and dunk their front feet to cool off.

There are also bottle-type waterers ($6) that have a metal "nipple" which the rabbits use to drink from. Rabbits that are unfamiliar with this system DO learn to use it quite quickly, and the water remains clean. No "tea" and no "cool-off dunking" with this waterer, though.

Breeding and Babies

The whole point of keeping rabbits is to get those little ones successfully into the world. If your animals are healthy to start with, you're more than halfway there.

Does and bucks should be at least six months old at their first breeding. Some will breed as young as five months, though.

A doe is caught and held by the scruff of her neck (NOT THE EARS!!), and lifted by supporting her with one hand on the scruff and another under her hips — watch out for those flying back feet with their sharp claws! Rabbits prefer to feel secure when carried, so you can tuck her under your arm (and wear a heavy jacket, just in case she decides to try to run!). Put the doe in with the buck — *never* put bucks into doe cages because they will fight. The buck will rapidly mount the doe and finish the job in a matter of seconds. The buck may fall off to one side and give a high-pitched squeaking sound — this is perfectly normal. That's it! The breeding is complete. Remove the doe back to her cage.

Four to eight hours later, repeat the breeding to be absolutely certain of success.

Write this date on your calendar or other record book, and then calculate 31-32 days ahead — that's when the bunnies are due. Occasionally, gestation can be as short as 28 days and as long as 38 days. During this month, make sure the doe has at least six ounces of feed per day, plenty of fresh water, and whatever treats you're inclined to provide.

On the 28th day, prepare the nest box. Make sure the box is clean and dry. Add to it a thick layer of hay, dry grass trimmings, leaves, sawdust, or other natural material. Put the nest box into the doe's cage, and scatter other nesting materials around her cage. If the doe is very close to giving birth, she will scurry around her cage, picking up the nesting materials in her mouth, and will move them into the nest. As the fateful moment approaches, the doe will begin to pull hair from her neck and chest, and line the nest with this soft fluffy fiber. Don't disturb her at this time — some sensitive rabbits may go to pieces if they are bothered now. Generally, when you see hair-pulling, the babies will be born within 24 hours.

Chapter Two
The "Next White Meat" — Rabbit

You'll be able to tell when the bunnies have arrived by taking a short and judicious peek into the nest box — distract the doe with a special treat so she doesn't get frightened. If the doe is vigorously stamping her feet, try another time. The newborns will be completely concealed by that fluff of hair. If you blow on the nest, the babies will wiggle around and you'll see the hair move.

The second day after the birth, the doe will have calmed sufficiently that you can inspect the bunnies. Distract the doe with another special treat, and carefully lift the top layer of hair. There, you will find a pileup of little rabbits which resemble tailless mice. They will wiggle and give little hops. Their eyes are closed, and they have very thin coats at this point. Put your hand in the box and feel around for any still, cold bunnies, and for any that might have been accidentally flattened by the doe — remove all these from the nest and dispose of them. Count the bunnies present and include that in your records for future reference. Cover the remaining bunnies with hair. Once a day, you can check again for any that might have died, and remove them. If you find any living bunnies outside of the box, go ahead and put them back — sometimes one will stay attached to a nipple and get dragged out when the doe leaves the nest. The doe won't pick them up as dogs and cats do with their young.

At about two weeks of age, the bunnies will start peering out of the nest box. By three weeks, most will have ventured out for a few minutes. If there are more than eight bunnies in the nest, you might consider cleaning and changing the box now. Return some of the old nest hair to the cleaned box, so it is familiar to the family.

By four weeks, the bunnies will crowd the feeder and be hopping all over the cage. They'll return to the nest in a hurry when startled. Make sure there is a constant supply of fresh food in the feeders.

At five weeks, you can separate the bunnies from the doe into a "growing cage," if you wish — although I think they do better if they are not separated until they are six weeks old. I like to keep the growing cage next to the doe's cage so the babies can touch noses with mama if they are so inclined. These growing bunnies should have all the feed they will eat.

Between five and eight weeks, the doe can be rebred. Actually, the doe can be rebred immediately after giving birth — these animals are remarkably fertile. But each litter will do better if they aren't competing for nutrients with another litter.

Harvesting the Garden

At eight weeks, the rabbits are ready to be prepared for the freezer. I prefer to do this process away from the sight and hearing of the other rabbits. The rabbit to be butchered is carried to the processing area. This consists of a table or flat surface, a piece of rope or twine, a collecting bucket, and a bucket half-filled with cold, clean water.

The rabbit is held on the table and stunned by hitting it sharply on the head between the eyes and ears. You may use a 1"-thick dowel or stick, a ball-peen hammer, or anything that does the job for you. I once saw a man do this with a "karate chop" to the rabbit's neck, quite successfully. Alternatively, you can hold the rabbit up by its back legs and just yank down hard on the animal's head while bending the neck

Chapter Two
The "Next White Meat" — Rabbit

upwards. These methods are very quick and, if done correctly, the animal "doesn't know what hit 'em."

The rabbit will now begin a brief period of kicking, which are reflex movements of the muscles — the rabbit is already dead. Remove the head. Tie the animal's rear legs to the twine and proceed to butcher the rabbit (see the Appendix for guidelines). The first butchering will take up to an hour or so — but after you've done a few, typically a single rabbit can be done in 15 minutes or so. If you really get into a rhythm, you can do ten an hour!

Save the skin for tanning if you wish. Place the guts in the collecting bucket, taking out the kidneys and liver (remove the little green sack, the gallbladder, *very carefully* by cutting off a bit of liver with it... if the gallbladder breaks, it will make the liver taste bitter), if you eat these delicacies.

Rinse the meat. Place the rabbit in another container, cut up into serving-size pieces, and cover with cold, fresh water. If your container is large enough, you can put all your rabbits into it as long as they can be covered with fresh water.

Young fryer rabbits can be cooked immediately, can be held in the refrigerator for up to 48 hours to "age," or can be frozen or processed into canning jars. Older rabbits need to be aged for the full 48 hours before use or freezing — or else they will be tough and chewy. Some recipes for rabbit are included at the end of this chapter.

Healthy Meat

Rabbits raised in clean cages and with plenty of wholesome foods are seldom sick. There are some common illnesses that you might see, though — but good hygiene will prevent these in the first place!

Snuffles: The rabbits get sneezy and appear to have a head cold. They look droopy, act listless, and sometimes will have a mucous discharge from their nose or eyes. Snuffles settles into a rabbitry when the animals are stressed, either by insufficient feeds, a buildup of manure and fumes, or from anxiety created by marauding predators, or even by severe weather changes.

Treat snuffles by cleaning up the environment, by giving the animals a teaspoon of apple cider vinegar in their drinking water — and, in severe and persistent cases, by dosing the animals' drinking water with antibiotics (which can be found in veterinary supply catalogs — see Resources). Keep in mind that if you have to give your animals antibiotics, *you* will get antibiotics when you eat them, and if you have to treat them this way continually, you're probably doing something wrong!

Coccidiosis: The sign of this condition is white spots on the animal's liver. *Do not eat a liver with white spots on it!!*

The rest of the meat can be eaten, though. Coccidia are an intestinal parasite that can cause persistent diarrhea, eventually leading to the animal's death. Like snuffles, it is a sign of a rabbitry lacking in simple hygiene — particularly if you have chickens that leave their "deposits" in the rabbit's food or water (chickens are coccidia carriers).

Coccidia must be treated with a coccidiostat, which can be purchased mail order from veterinary supply catalogs. If one animal in your herd has coccidiosis, treat them all.

Chapter Two
The "Next White Meat" — Rabbit

Ear Mites: These nasty little invisible critters burrow into the inside of the rabbit's ear canal, and set up an infection that results in a crusty yellow crud in the rabbit's ears. This condition is very common, even in otherwise clean herds. Left untreated, the animal's ears will become clogged with this itchy crust, and the rabbit may eventually die from infection.

Look into each rabbit's ears once a month. If you see the crusty patches, pour a few drops of ordinary cooking oil into the rabbit's ears using an eyedropper or squeeze bottle. The oil covers the mites and smothers them, while causing the rabbit no discomfort (other than giving them a greasy ear and neck!). Treat all the rabbits in your herd once every two weeks for three months, and you will virtually eliminate ear mites from your property.

Unwilling to Breed: Bucks five months and older should immediately get to work when presented with a doe. If a buck doesn't actively try to breed under these conditions, he should be culled (put in the freezer).

Does will sometimes be uncooperative, though. Try her with another buck if you have extras — sometimes she just feels picky. Try her several times in one day. If that doesn't work, put her in a cage right next to the buck so they can touch noses and get acquainted. With sudden weather changes, does will sometimes be uncooperative — also, during very hot weather they may be disinclined to breed. If the doe is consistently hard to breed, you might consider replacing her with a willing breeder.

Bunny Killing: Does will occasionally kill a single bunny in a litter by stepping on it when getting in and out of the nest.

This is sad, but normal — usually, these will be flattened somewhat.

Does that are undernourished or eating a diet severely lacking in protein, though, may purposefully eat all or some of their litter. Generally, these bunnies will be eaten starting at their feet — so if you see chewed legs on babies in the nest (and if there are no rats bothering the herd), it's probably the doe who is responsible. In well-fed does, bunny-eating MAY take place ONCE if the rabbit is badly frightened by something. But well-fed does who eat their babies on more than one occasion should be culled immediately. Also, don't save any of her offspring for breeding, since this bizarre behavior may be transmitted in family lines.

Hair Pulling: Usually, if you see hair pulling going on, it will be in a litter where one of the bunnies is responsible for yanking tufts of hair out of another bunny's back. Occasionally, if this isn't discouraged, the result will be sores on the victim's back from repeated attacks.

There is some indication that this may be due to a protein deficiency, or to a lack of fiber in the diet, or possibly even to boredom. Try increasing the feed to 18% protein, putting some pieces of fruitwoods (apple and pear are best — but don't use cherrywoods because they can be toxic in large quantities) or a piece of a pine 2 x 4 in the cage for the rabbits to chew on, or even fresh grass trimmings for them to pick at. Some rabbits enjoy having an aluminum soda pop can (pull-tab removed) in their cage to carry and flip around.

If the hairpulling continues, cull the offender — you can eat small rabbits, too.

Chapter Two
The "Next White Meat" — Rabbit

If it is the mother rabbit who is responsible, cull her after the young ones are four weeks old.

Sore Hocks: The long underside of a rabbit's rear feet are usually covered by a thick layer of fur. Under constant pressure from a wire cage — particularly a wire cage that hasn't been cleaned in a while — those hocks can develop circular sore spots, where the hair wears off and an actual break in the skin occurs. As you can imagine, these hurt and make the rabbits unhappy and unwilling to do their jobs.

The cage should be thoroughly cleaned and bleached, scraping all waste deposits from the wire. The rabbit's feet should also be cleaned, although this is easier said than done, and the sore should be treated with a good antibacterial spray for livestock (see vet catalogs under Resources). To ease the pressure on these sores while they are healing, a flat board or several thickness' of cardboard can be put into the cage as a small "platform" (that is, don't cover the whole cage floor), so the rabbit can rest his hocks from the wire. Given this treatment and attention to cleanliness, the sores should heal within ten days.

Rabbits that have sore hocks which refuse to heal with this treatment probably should be culled.

* * *

For general good health maintenance, many long-time breeders recommend the use of apple cider vinegar, 1 teaspoon per quart of drinking water. You can give this to the rabbits day in and day out. Rabbits given this simple treatment often appear livelier and have better coats than rabbits that don't get it. It may supply vital nutrients that simply can't be gotten any other way.

Rabbits also enjoy having something hard to gnaw on, such as the previously mentioned hunk of pine 2 x 4. The wood shouldn't be treated — just a plain, clean piece of wood.

Any hard root vegetable makes a pleasant treat for rabbits — potatoes, carrots, beets, mangels, — and fresh fruits of all kinds are relished, including sliced apples, oranges, and grapes. Lettuce, cabbage, even onion tops are good for rabbits... any food you can eat, your rabbits can eat in its raw state, too. They also enjoy herbs of various types, including comfrey and wild "lamb's quarters." Just remember that treats shouldn't make up a major part of the animal's diet.

For the excruciatingly thrifty among us, supermarkets routinely toss out green produce which has wilted or passed its prime. A few enterprising rabbit owners have arranged to collect these throwaways and make them a regular part of their animals' diets. This is excellent and very cost-conscious — but make sure they get their pellets, too, since they need the protein!

Savoring the Harvest

Ah! Now to the important part! Rabbit meat is naturally very lean. For this reason, the chef needs to have a light touch when cooking and use additional liquids when preparing these meats — young, tender fryers can be cooked to leather-toughness by letting them go too long under high heat; and older "broiler" rabbits benefit by a par-boiling before being cooked by other methods. Try substituting rabbit in your favorite chicken recipes.

Chapter Two
The "Next White Meat" — Rabbit

Here are three of our favorite rabbit recipes:

RABBIT POT PIE
(serves 4-6)

1 fryer rabbit
4 cups water
3 peeled and sliced carrots
1 cup chopped celery
2 medium onions, chopped (tops included)
3 tablespoons butter or cooking oil
3 tablespoons flour
1 teaspoon salt
1 pie shell to fit your pan, top and bottom

Combine rabbit, water, half the salt, and the carrots and bring to a boil. Cook for 20 minutes. While cooking, lightly sauté the celery and onions in the oil. Strain the rabbit and carrots from the broth (saving broth), and debone the rabbit, cutting the meat into bite-sized pieces. Use 2 cups of broth, stir in the flour, and heat this until thick and bubbly. Add to this the rabbit, vegetables, and remaining salt. Pour this fragrant filling into the pie shell, cover with the remaining pastry, cut slits in the top, and bake at 350 degrees for about 50 minutes. If the crust gets too dark, cover with aluminum foil.

Let this sit for 20 minutes to cool slightly after removing from oven. Fabulous!

BEER RABBIT
(serves 2 really hungry guys, or a whole family)

2 fryer rabbits, cut into serving pieces
6 slices of diced bacon
2 cans dark beer, opened for two or three hours beforehand
1 bay leaf, crushed
2 chopped onions
4 tablespoons prepared mustard
½ cup flour
1 teaspoon dried thyme
2 teaspoons minced fresh parsley
1 cup whipping cream or half-and-half

In a large fryer or Dutch oven, fry the chopped bacon until slightly transparent, then add the onion and cook together. Remove when the onions are transparent and the bacon is cooked crisp. Flour the rabbit pieces, and brown in the bacon fat, turning frequently to prevent burning (but if they burn a little, it just improves the flavor!). In a separate bowl, mix all the other ingredients together, and pour over the rabbit pieces in the pan. Cover and simmer or oven bake at 300 degrees for about 40 minutes.

If you think these are good the first day, you should try them cold the day after (if there're any leftovers)!!

RABBIT SAUSAGE
(makes five pounds)

4 pounds raw boned rabbit
1 pound fatty pork butt or shoulder

Chapter Two
The "Next White Meat" — Rabbit

2 teaspoons salt
1 teaspoon each ginger, mace, cumin powder, nutmeg
1 tablespoon each ground black pepper, sage powder

Grind the rabbit and pork together, then thoroughly mix in the spices and seasonings. Refrigerate or freeze until use — then, fry as patties or crumbled like regular sausage. If you prefer a low-fat sausage, reduce the pork to ½ pound, and increase the rabbit by ½ pound.

Chapter Three
The Portable Egg Factory — Chickens

Even if you've never seen one "in person," you already have some pretty clear mental pictures of what a chicken is supposed to be. For some people, a chicken is a sleek, white-feathered, strutting bird with a bright red comb — and for others a chicken is a chunky, brownish-red feathered bird, clucking to a bevy of fuzzy chicks in a barnyard.

Well, of course, these are both representative of different breeds of chickens. In fact, there are dozens and dozens of breeds, each with a unique "look," body style and carriage, and with different egg-laying capabilities. History's original chicken-type bird probably laid around 50 eggs per year — which made an egg a pretty special commodity. Today's breeds, ones bred especially for their egg-laying capacity, can give around 300 eggs annually, about 25 dozen!

Chickens can be divided roughly into four main groups, and one subgroup. These are: (1) egg layers; (2) meat producers; (3) egg and meat producers; (4) fancies; and (5) bantams.

This handsome rooster is a "Turken" or "Naked Neck," a large chicken noted for its lack of neck feathers.

The Breeds and How to Get Them

Chickens bred specifically for eggs tend to be physically smaller birds, with hens reaching four to five pounds at maturity. Typically, these are Leghorns or Leghorn-like (pronounced "leggern"), sleek, white-feathered birds. Their behavior is fast-moving, inquisitive, and querulous. These hens can lay over 300 white eggs annually; the majority will be "large" size, with a few medium or small eggs at the early stages. If you keep these hens for several years, their annual output will drop in numbers but the size of their eggs tend to increase with the chicken's age, so three and four year olds might give 150 "jumbo" eggs annually. At this point, the hens

Chapter Three
The Portable Egg Factory — Chickens

"molt" or lose their feathers, and stop laying for a period of about two months. After regrowing their feathers, they'll start laying again, with a slight decrease in total output. These hens lay well until they are about three years old, and then they tend to poop out — most "egg factories" send these chickens to the soup factory at their first molt. Hens refuse to "set" or hatch eggs — they just don't understand the concept. Roosters of these breeds are small, feisty and usually end up as small fryers at four months.

Strictly meat-type breeds are the Cornish and Cornish-White Rock crosses (sold at six to eight weeks as the delicacy "Cornish Rock Hens"). These are really chunky tank-like birds, reaching eight to 12 pounds, and sometimes more, with a high percentage of breast meat. They grow fast, but require a feed with a higher protein percentage to encourage growth of their heavy bones. Some catalogs offer these as "barbecue roasters." If you strictly are looking for meat, better birds can't be found anywhere — but don't expect them to double as layers... they produce a measly couple dozen eggs annually. Hens don't set.

Dual-purpose birds include most of the "heavy-bodied layers," or more-traditional chickens that you'd expect to see wandering around an old homestead. Breeds such as the Rhode Island Red, Brahmas, Orpingtons, and the Rocks, all reach a large meaty size of about ten pounds. They are fairly slow growing, relatively slow moving birds, and usually hens give about 200-250 brown or off-white large eggs yearly. Hens of some of these breeds will set, although they are sometimes casual about their motherly duties. Roosters can reach 16 pounds at maturity, and have a loud crow that will echo around your neighborhood.

Fancy breeds, such as the Polish, Silver-Spangled Hamburg, and the Houdan, are birds that have been bred over time to look really neat — bright sparkling feathers in unusual colors, or with spots, or even with a little tuft of feathers on their heads as a "top hat." Generally these birds are small to medium-sized, poor to moderate layers, and non-setters. Fighting "game" birds, with their sleek build and nasty tempers, can fit in this class, too. The fancies' main reason for existing is that they're interesting to look at. They'll lay around 100 eggs yearly.

Bantam chickens are scaled down or miniaturized models of the regular larger breeds — or maybe the big birds are the product of "breeding up" bantams. They come in the same breeds as larger chickens, but typically don't get bigger than four to six pounds. Banties begin laying around four months of age and produce about 200-250 small to medium brown eggs yearly. In behavior, they are fast-moving, quick to fly, and nervous.

They are probably the best-setting birds that exist. They are dedicated mothers, too. In fact, these birds are so desperate to set eggs, that you really have to discourage them from "going broody" by continually removing ALL eggs and egg-shaped things from their nests. They'll keep cranking out eggs as long as they're not setting — but our banties tend to lay off egg production during the coldest part of winter, no matter what. Banties will set ANYBODY's eggs, too — they'll hatch geese, ducks, and regular-sized chicken eggs. If you're looking for an "automatic incubator," bantams are the way to go.

Chicks of any of these types can be purchased from large hatcheries through the mail (see Resources). Day-old chicks are shipped via airfreight overnight or within two days almost anywhere in the continental US. The chicks are able to tolerate this, because the shipping boxes are fairly well insulated, and the chicks absorb the remnants of the yolk just before they hatch — which provides them with a 72-hour "food" supply that tides them over this period. If you wish to buy chicks of specific breeds, they'll run about 40 to 90 cents each — but you can often find "bargains" and "specials" of egg-layers, "heavy breeds," fancies, or meat-type birds for $9 to $20 for 25 chicks. The "unsexed" (mixed male and female chicks) and "cockerels" (baby roosters) are usually the least expensive; there may be a 25-chick minimum order.

You may also be able to find adult birds locally, at pet stores or from private individuals in your area. Adult birds may run $3 to $10 each, or higher for fancy breeds.

Arrival and Housing

Chicks arrive with your mail, or you pick them up at your local post office. You should have a pen prepared for the chicks, with a "water fountain" and "chick-starter" feed in a low feeder and scattered over the flooring.

Chickstarter is a special, finely ground feed mix with a high protein level. Chicks cannot eat coarsely ground feeds that would be suitable for older birds. In the Midwest, it costs $7 per fifty pounds — double that price on the Coasts. Chicks don't eat very much to start with, so 50 pounds should last until two dozen chicks graduate to regular feeds.

Water fountains can be made from a plastic screw-on trough and a mayonnaise-type jar (see illustration). The screw-on trough costs $2-$3 new if plastic, a little higher if metal. For the first two days, chicks will pick everywhere and won't use the feeder much, but having one will keep your grain clean and prepare the chickens for using feeders later in life. A chick feeder consists of a screw-on trough with holes ($2-$4), and another mayonnaise-type jar.

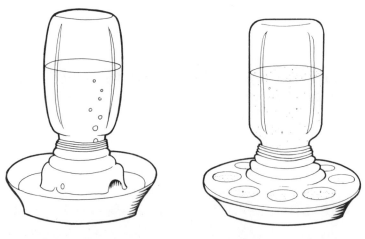

Chicken waterer.   *Chicken grain feeder.*

Add a tablespoon of sugar to the waterer — or alternatively, use "chick electrolyte" powder in the water. This simply provides a little glucose energy boost for the chicks, to help them get started. It's not critical for healthy chicks, but it can help weak chicks survive the stresses of their trip.

When you place the chicks in their new residence, pick each chick up individually, dunk the tip of its beak in the wa-

Chapter Three
The Portable Egg Factory — Chickens

terer (don't dunk beyond the tiny nostril-holes in the beak!). Set the chick down and it will tilt its head back and swallow its first mouthful of liquid.

If you have a "brooder" (new, about $300), a set of stacked cages with a heating unit, designed specifically for chicks, by all means use it now! Turn the unit on, adjust the temperature to 95 degrees, fill the feeders and waterers. When the interior is warm, put the chicks in. That's it!

I've been buying chicks since 1980, and every year I think I'm going to get a brooder — but I always seem to end up with the homemade version described here. Take two cardboard boxes and put the smaller of the two inside the larger — creating a double-walled single box. The interior of this box should be about two feet by two feet, roughly. Lay a clean, old towel in the bottom of the box to provide traction for little chick feet. Poke about a dozen small holes from inside to outside the box about two inches up from the bottom — this provides for fresh air flow at the chick's level. Put the feeder and waterer into the box, but separate them so that the chicks need to walk a little to get at them. Scatter feed on the towel — new chicks will "discover" their food right at their feet. Now, carefully attach a clip-on lamp with a 60-watt bulb ($7-$11) to the side of the box so that the lamp is inside about four to five inches above the box floor, on the side of the box AWAY from the food and water. If you have an incubator thermometer ($3), put it on the floor beneath the lamp. It should warm up to around 100 degrees. This warm place will be your "hen-substitute" and needs to be on constantly day and night for the next two weeks, at least. (If you don't have electricity for this, you may place a half-gallon plastic water jug, filled with hot water, into the "brooder," but you have to change this often enough that it doesn't get below 90 degrees — or your chicks will die.)

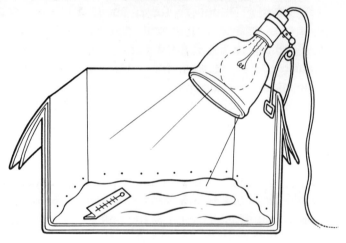

Side view, homemade brooder box.

Dunk beaks, put chicks under the lamp to warm up.

You'll notice that your new chicks were cheeping loudly when they arrived — they were cold and scared. Placed under the heat, they will gradually quiet down. Some will fall asleep for a little while, and a few will begin pecking and scratching right away. If chicks pile up under the lamp, they are very cold — you might lower the lamp a little (an inch or so) to provide a wider area of warmth so the chicks don't have to pile up. Chicks at the bottom of the pile can get mashed, so piling should be discouraged. If necessary, add another lamp to the box for extra warmth.

A word of caution: If you're using lamps, BE CERTAIN they cannot touch the sides of the box, or come loose and fall

Chapter Three
The Portable Egg Factory — Chickens

into the box!! The last thing you need is to burn down your chicken brooder — or your home!

The chicks can stay in this box for about two weeks, provided they don't get too crowded. Remove the towel after two days (you'll want to!), and put newspaper, hay, sawdust, or other natural material on the floor (remember the caution about the lamps!). Put a handful of clean garden dirt in with them, too. Every two days, raise the lamps up an inch or two — this gradual cooling allows the chicks to adapt to outside temperatures. If they chirp piteously after you've moved the lamp up, they're too cold and you may need to lower the lamp for another day or so.

Graduate the chicks up to a larger box after two to three weeks. You may have to move them to larger boxes several times until they have grown regular feathers, "feathered out." With feathers and decent non-freezing weather, the chicks can be graduated to unheated cages or to your outside run.

Chickens can be kept in cages, allowing at minimum one square foot of floor space per bird. Using this calculation, in a 24" x 36" cage (the same double decker shown for rabbits on page 14), six birds could live fairly comfortably. Personally, I'd only put four or five birds in an area that size, just so that they'd be that much more comfortable. If your cages are portable, you can move them around and spread manure "automatically." Cage chickens need fresh greens everyday (lettuce trimmings, cabbage, carrot tops, etc.), plus a small box with plain old clean dirt in it... they do get vital nutrients from the occasional mouthful. Also, your cage birds will seem more contented if they have a short time every day to scratch on the ground for bugs.

Backyard Meat Production

Chicken pens should allow 18 square inches for each bird, be surrounded by wire on all sides, and have a stretched wire roof (to prevent predators from swooping or climbing in). Make this tall enough that you can walk underneath it, too. There should also be at least one shady area for the birds to get out of the hot sun. Feeders and waterers need to be changed every day. The chickens will soon peck all the greenery out of the area, so you'll need to provide them with some kind of green feed along with their regular rations.

The birds will want a "roosting" area, consisting of raised and rounded poles or tree branches to perch upon during rest and at night. The chickens will seek the highest possible place for their rest, perhaps as a throwback to their wild ways when they needed to sleep out of the reach of roaming predators. These poles or branches need to be secured and braced, since a dozen fat hens can weigh as much as a person does!

Winter housing for chickens doesn't need to be fancy. Add extra organic flooring materials (straw, sawdust, pine shavings, etc.), and block drafts into the pen or cage. An unheated garage or basement is ideal, as long as it can be easily aired on warmer days, and cleaned routinely.

Feeding

Just like with rabbits, agriculture has provided us with "perfect" chicken feeds, already mixed and measured to give the best nutrition to our birds.

Chick starter, mentioned previously, is ideal for baby birds. If you don't have access to this feed, you may substitute corn meal soaked in whole milk, changing it as it sours.

Chapter Three
The Portable Egg Factory — Chickens

Chicken feed comes as "mash," which is loose grain, or as pellets, about $5 for 50 pounds in the Midwest. Both will fit into your feeder trays, or can be placed directly on clean ground. Chickens will waste less feed if it is put into trays, though.

Birds enjoy scratching at the ground, and pecking at tiny objects. Even birds in cages will show scratching behaviors. Chickens also will lay down, flick their wings, and throw loose soil under their feathers — this is a "dirt bath," and helps to rid the birds of lice and other parasites — so don't be shocked when you see this happening!

There are stories from the Depression era, of farmers who ran out of grains and had to hunt up feeds for their hungry chickens — supposedly, road-killed critters were scraped up and hauled to chicken pens...an excellent source of protein. Today, I'd be a little worried about bringing potentially disease-ridden dead things onto my property!

But chickens are omnivores — they can eat nearly anything we can. The addition of milk to a bird's diet will bring immediate and amazing improvements in health and feather condition. Along with their daily grain ration — which can be as simple as whole corn, if you're supplementing with lots of other stuff — you can feed fruits, chopped vegetables, all kinds of greenery, your own leftovers, cooked meats and guts from other animals, and even used cooking oils or other cooked fats such as bacon grease. Your chickens will really benefit, too, if you wash and crush egg shells and include that in their diet... make sure they are crushed to the point that they no longer resemble egg shells, though — you don't want to train your birds to eat eggs! The one thing you don't want to feed your birds is: birds — too much chance of spreading disease.

If you are raising chickens strictly to eat, you can begin butchering when they reach two pounds — for Cornish types, this can happen at three to four weeks of age and gives you those tiny delicacy "game hens." For other breeds, butcher when they reach four to five pounds for a nice frying bird, or up to eight to nine pounds for a "roaster."

Eggs, Nests, and the Next Generation

Young hens begin laying between four and one-half to six months of age. Roosters start to cock-a-doodle-do at about the same age. If you intend to raise more chickens from your birds, keep one rooster per ten or twelve hens. The rooster you save should be the largest and most attractive of the batch, and the one with the most interest in hens. Roosters have larger combs than hens, tend to hold their bigger tails more upright than hens do, and tend to be a little more aggressive. Their colors are brighter, too. But if you're not interested in raising chickens, the hens will still lay good eggs without ever setting eyes on a rooster.

Your first eggs may be quite small, may lack a true shell, or may have streaks of blood on them — all perfectly normal but weird. Before long, you'll be getting almost as many eggs every day as you have hens. They'll tend to lay in the morning hours, so by lunch time your daily eggs will be ready. Fertile eggs will have a tiny "blood spot" in the yolk, which is the sign of fertility. This won't affect the flavor at all and will become unrecognizable when cooked, but if it grosses you out, you can remove it when you use the eggs.

Chapter Three
The Portable Egg Factory — Chickens

Some hens will lay eggs wherever they happen to be standing at the time. Others are very specific, and will hunt out dark corners or hidden spots under bushes. A nest box — essentially a 12″ x 12″ x 12″ box with a little hay or dirt in the bottom — gives hens a private darkish area to put their eggs. You can construct a line of nest boxes if you have several chickens who might be laying at the same time — although the hens will tend to prefer nests with an egg in them already. Fake plastic eggs, or a single hard-boiled egg (with an "x" on it so you can tell it apart), will often encourage young hens to leave their eggs where you can find them.

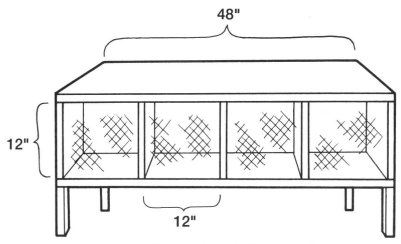

Chicken nest boxes — line with hay or sawdust.

Unless you and your family are egg-fiends, you will be overwhelmed by fresh eggs in no time. After you've swamped your friends and neighbors, you can preserve extra eggs for later use with recipes at the end of this chapter.

Backyard Meat Production

If you intend to hatch eggs, wait until the hens have been laying for a month before saving any for hatching. You'll have to keep a rooster in with the girls, as well. If you have any Bantam hens, they will take care of the hatching as the spirit moves them.

Lacking banties, you will need to collect eggs daily, and store them in egg cartons. Turn the eggs every day until you have a sufficient supply for your needs — I'd recommend no more than 25 to start will. You'll need an incubator, too. The least expensive model which is effective costs about $30 new. It's made from a Styrofoam box, and includes a heating element, thermometer, and a wire floor upon which to place the eggs. These come with instructions on how to incubate the eggs.

Basic incubation means giving the eggs a warm environment of 101-103 degrees, moisture as humidity in the incubator air, and turning the eggs at least four times daily (six is better). The incubator needs to be prewarmed. Mark the eggs with an "x" on one side and an "o" on the other. Lay the eggs so that all "x"'s are showing. Leave them in the warm incubator, with a little water in the internal tray, for two days without turning. On the third through 19th day, turn four to six times daily (including during nighttime hours). Turn from "x" to "o," then back at the next turn. After the 19th day, don't turn anymore... they'll start hatching out on the 20th to 22nd day. Leave the chicks in the incubator until most of the batch has hatched — then, treat them just like "new arrivals" with a brooder box.

You can put together a "home incubator" using a pair of cardboard boxes, a lamp, and a dish of warm water to humidify the air. An incubator thermometer placed at the level of

Chapter Three
The Portable Egg Factory — Chickens

the eggs helps you adjust the temperature. These need to be turned and rotated under the light AT LEAST six times a day; if any eggs get cool to the touch, they'll probably fail to hatch.

It is VERY RARE for all incubated eggs to hatch. Home incubators like this one, and commercial Styrofoam ones, will give you about 50% hatch and survival. If you can get 70% hatches or better, you're doing really well. An attachment for the Styrofoam incubator, an egg turner ($30 more), makes the job a lot easier on you — no more midnight turning. Expensive $300-$600 incubators give a better hatch rate, but you'd *really* have to want to raise chickens to make this a worthwhile investment.

Healthy Birds

Chickens are subject to a few conditions that impair their health — generally speaking, any bird that appears sick and listless should get only one specific treatment: the ax. This prevents the spread of disease to the rest of your flock.

Pecking each other: Chickens develop a "pecking order," with one hen being the dominant bird. She can peck on everybody. The next hen down in the order can peck everybody except the bird "above" her... and so on down until the bottom bird, who can be pecked by everybody, and can peck no one in return. This bird is "the bottom of the pecking order." When chickens are overcrowded, bored, or nutritionally deprived, they can really go to work on that bottom bird — and a single spot of blood seems to drive them into a pecking frenzy. The victim can be pecked to death. As a remedy for this, you can enlarge your confinement area, or reduce the number of birds within the area; you can increase greens feeding, or provide a higher protein feed, or add more protein foods to their diet; or you can find another source of amusement for the birds — a bright red soda can suspended at their eye level on a string can be very interesting to chickens. The victim of the pecking order can also be treated by putting a bit of pine tar (see Resources for veterinary suppliers) on small wounds — it tastes bad to other birds and discourages them. If she has been badly injured, I'd suggest culling.

If chicks are pecking each other, particularly if they have pecked each other's toes raw, they are overcrowded AND underfed. Commercial poultry raisers "debeak" chicks — clip off half of their upper beak with something like toenail clippers — to prevent this problem. But, our backyard flock will be happier and healthier if we just give them more space and better meals. Adding milk to chick feed seems to prevent this, too.

Lice: As horrible as it sounds, poultry of all kinds will become walking homes to lice — these tiny pale oblong critters tend to be found under the "armpits," around the "vent" (anus), and on any exposed skin. Chickens badly infested with lice can actually become anemic from the bugs' constant feasting. Plus, these critters can also live on US!

Fortunately, the remedy is simple: lice powder, or rotenone garden powder, or even garden diatomaceous earth can be used to "dust" the birds. If you find lice, then immediately dust each chicken in your flock by holding the birds upside down by their feet. Shake the dust into the bird's feathers, ruffling the feathers so dust will penetrate to the skin. That's it. Put dust on their nighttime roosts, and in their dust-bath

Chapter Three
The Portable Egg Factory — Chickens

holes. Do this every two weeks for three months, and the lice should be eliminated entirely.

Egg eating: A very bad habit, indeed. There are different opinions on why this takes place — some authorities suggest that overcrowding and lack of protein in the diet (once again) contributes; others think it only happens when an egg is already cracked and the seeping contents bring on a feeding frenzy. Personally, I've observed that it is only one or two hens who will engage in this, and occasionally a rooster. The only sure-fire remedy, as far as I can tell, is the ax.

"Missing Head Syndrome": You come out in the morning after a normal enough night, and find two or three hens laying in their cage or pen, but their heads are gone! Is this another strange case of space alien "chicken mutilations"???

Probably not. Most likely, you'll find signs of a struggle — feathers scattered around, other birds acting a little nervous, and the heads either missing or tucked into crooks of nearby tree limbs. This is the *modus operandi* of two evil villains: opossums or raccoons. Opossums are sloppy in their deadly work, and frequently chew up the leftover bodies as well. Raccoons are wasteful predators — they'll often rip heads off of their victims, and leave the rest behind. We once had seven young ducks lose their heads in a single night to ONE raccoon!

Realistically, the only permanent solution is to eliminate the offending predator. Humane traps can be baited with fresh liver and placed near the likely routes of entry to the chicken pen — but, I will caution you not to ask for the assistance of your local Humane Society. They may have uncomfortable questions and comments about your meat garden! If you're of a mind to, both 'possums and 'coons are perfectly edible — but, please, please, if there is rabies in your area, don't risk eating these critters! Handle even the dead predators with extreme care, and dispose of corpses where other animals won't be able to get at them.

Prolapse: A hen looks as though she is straining to lay one really big egg. When you check on her, it looks as though a half-laid cantaloupe is trying to be born — this is a prolapse, or "fallen gut," which has protruded from her "vent." Sadly, the only remedy is, once again, the ax.

Burned Comb: You'll see this condition — a blackened, dying section on a bird's comb — after periods of extremely cold weather. Basically, burned comb is a bad case of frostbite. Unless the damage is extensive and extends into the head skin, this will heal without any treatment — but the bird will lose a bit of its comb. If you see burned comb, you need to improve the winter housing for the birds, which may be as simple as putting flattened cardboard boxes up around the pen or cage to block cold winds.

Harvesting the Chicken Garden

When the birds have reached the size you'd like, here's the plan: the night before you want to do the job, remove all food but leave plenty of water. Confine the birds, so they're easy to get at.

The morning of the butchering, prepare your work area: a hatchet is handy but not necessary, a pair of sharp knives, a flat surface or table, and a very large pan (a "water-bath can-

Chapter Three
The Portable Egg Factory — Chickens

ner" is ideal) filled with nearly boiling water. Keep a bucket nearby for wastes.

Bring your meal-to-be to the butchering area. You can chop the chicken's head off — hold the wings and legs with one hand, and stretch the neck and head out with the other. Lay the head onto a wooden board or chopping block, and strike the hatchet on the neck between the head and shoulders. If you've done this well, the head will be completely severed. Alternatively, you can grab the chicken by the head and swing it sharply around your head in a "crack-the-whip" motion — this will snap the neck (and may rip the head off, too).

Immediately after you remove its head, the chicken will begin hopping and flapping its wings. These reflex movements are the source of the old saying, "Acting like a chicken with its head cut off"... that is, acting pretty much out-of-control. These random and undirected movements do help to pump blood out of the chicken's body, and make the meat keep better, so let them happen. It should be over within a minute.

Take the chicken and dunk it very quickly into the nearly boiling water. This loosens the feathers so that you can quickly strip them off. Don't leave the bird to soak in the water — 30 seconds should be sufficient — otherwise the skin will tear when you try to pluck. You can also "dry pluck," which means picking all the feathers off by yanking. Either way, the bird should be completely stripped of feathers. If you don't like the skins on your meat, you can "peel" the bird in the same way you'd skin a rabbit.

Now, you can go ahead with the butchering, as indicated in the Appendix.

Saving the Harvest

Home gardened eggs are so much more flavorful and colorful than supermarket eggs, it may come as a real shock. But the difference between home gardened chicken meat and that *stuff* they sell at stores is so vast, you may never be able to go back to the soggy store *stuff*. It probably won't surprise you that your home gardened chicken does cost a little more than the supermarket variety, perhaps as much as 20% higher — but after you've eaten some of this meat, you'll believe that the improvement is worth the small extra cost. Besides, if you're raising rabbits, too, the costs will balance out and your total overall cost will STILL be lower than market costs.

Chickens should be "aged" by holding them in your refrigerator in a plastic bag or other covered container for 48 hours before eating them — otherwise, they will be tough and leathery. After the aging period, use them as you like in any of your favorite recipes.

Eggs mount up very quickly. What do you do with the extras?
You can freeze eggs, quite readily, following this recipe:
Crack a dozen to 14 eggs into a bowl. Remove bits of shell and blood spots if you like. Carefully beat these eggs together with a fork, avoiding adding extra air to this mix by too vigorous beating, or else your eggs will be tough when thawed. When these are mixed, you can add EITHER ¼ teaspoon salt per each three eggs used OR ¾ teaspoon sugar per three

Chapter Three
The Portable Egg Factory — Chickens

eggs. Salt should be added if you'll use the eggs in regular cooking; and sugar should be added if you'll use the eggs in cakes or cookies. I usually freeze my eggs without either, and they come out okay.

Now, pour the eggs into an ice-cube tray that has been lightly oiled or sprayed with non-stick pan coating. Freeze. Remove from ice tray and store in a gallon-size freezer bag. Each "egg cube" will equal one medium-sized egg. When you plan to use one, take it out and let it thaw completely before use — then, it's just like a freshly beaten egg!

If you prefer, you can separate yolks from whites before freezing for special recipes. Don't beat or mix the whites. Yolks can be lightly beaten. Mark containers carefully, indicating how many whites or yolks are enclosed. Thaw completely before using.

Pickled eggs, for some reason I can't figure out, have become as costly as some gourmet foods — and they are so easy to make that it's ridiculous. Here's how: use eggs that were laid a week to ten days earlier so that they are easy to shell. For each mayonnaise-type jar, take 13 medium-sized eggs and boil for ten minutes. Quickly dunk in cold water, then shell. While you are shelling these eggs, heat three and one-half cups white vinegar and one-half cup water. Add to this a thinly sliced onion and a pinch of pickling spices if you like.

Put the shelled eggs into a thoroughly washed mayonnaise-type jar, and pour the boiling vinegar solution over them. Quickly cap the jar (use a canning lid and band, if you have them). Let this cool off, and then refrigerate. These will keep for months and months — slice some onto a home-grown ham sandwich for a heavenly taste!

Chapter Four
The "Wild Things" — Quail, Pheasant, Guineas, Ducks

There is a special and unique "gamy" flavor in the meats of wild birds — a richness which is simply unequaled in domestic birds. These fowl lend themselves to fancy occasions, memorable events, and times when "only the best will do"... and these home-raised birds will be some of the finest gourmet eating you will ever enjoy.

Selecting for Ease of Care and for Flavor

Each of the game-bird varieties that will be covered in this chapter have care requirements that are quite similar to chickens. Except for Coturnix Quail, cages can be the same ones used for chickens, the feed will be similar to chick starter, and the growth period is roughly the same. A primary difference between game birds and domestic ones, though, is that game birds lay many fewer eggs — they just don't produce enough to make raising them for meat AND eggs worthwhile. (The exception here is the Coturnix, as you'll see shortly.) However, there is sufficient egg production so that incubating and raising MORE gamebirds is more than sufficient.

Coturnix Quail

There is probably no more rewarding small game bird than this amazing quail. The Coturnix generally weighs no more than eight ounces at maturity, about the size of a dove. Dressed out, each complete bird weighs only about three and one-half ounces. It takes at least two to make a meal — and for hungry folks, six birds on a plate won't be too much. This looks and smells like a medieval feast!

The truly remarkable things about Coturnix are their reproduction and growth rates. Coturnix chicks hatch in 18 days. The eggs are about the size of a large grape, brown and dark brown splotched. Newly hatched chicks are slightly larger than a quarter, and full of life and energy. Like chickens, the teeny babies begin scratching and pecking shortly after they stand up. Within a week, you'll see feathers poking out of their baby fluff, and after four weeks or so they will be feathered out completely.

These birds require cages that have smaller wire mesh than chicken cages — ½" x ½" mesh is ideal. Additionally, quail cages shouldn't be more than eight inches tall, otherwise the birds try to fly and will literally scalp themselves in the process. Generally, if you wish to produce fertile eggs for incubation, one male is kept in an 8" x 8" x 10" cage with two or three females. Cage "banks" can be purchased from catalogs just for these birds, so you can stack their living quarters five and six levels deep. Collected eggs should be kept at room temperature until you have enough to start the incubator,

Chapter Four
The "Wild Things" — Quail, Pheasant, Guineas, Ducks

turning the eggs daily in the meantime. Coturnix won't set eggs. I wouldn't leave hatching even to a banty, since the chicks are so tiny and fragile.

This small quail cage is three feet high and only twelve inches deep. Each level is only ten inches high, and consists of "group pens" with mixed male and female birds. Cage height here is 2" taller than the recommended 8".

Males and females are easy to identify — at about six weeks of age, the males develop a rounded bulge at the vent, which will extrude a "shaving cream" foam when pressed.

Backyard Meat Production

They make a sort of yodeling or cackling call that is quite loud considering their size. By eight weeks of age, the little quail are full-sized and laying eggs.

They'll lay year around if the temperature stays above 65 degrees. They must have light at least 12-14 hours each day and have plenty of good food and fresh water. Five Coturnix eggs equal one chicken egg, if you decide to cook with them. The eggs are ideal for pickling, and make wonderful, exotic holiday gifts.

Coturnix require high-protein feeds. They can do quite well on chick starter rations, but will do a tad bit better if you can find "game-bird feed" in your area. For chicks, this should be finely ground in your kitchen blender so that it is easy for the youngsters to consume it. Older birds eat the crumbled feed without any difficulty. Both food and water should be offered using "quail" feeders and waterers — these have smaller holes and troughs than chicken equipment, so you're less likely to lose a little bird to drowning or choking. Additionally, the eggs will have better color and the birds will be healthier if they have access to greens and raw vegetable bits on a daily basis.

When butchering these birds, simply snip the head off with scissors and dry pluck the body after it has stopped flapping. (You can skin them, if you prefer.) Remove the internal organs by cutting the bird down its back — this makes it look more attractive when served breast-up on the plate.

Roast these guys in a group at no more than 325 degrees, for 30-45 minutes. Baste with butter, gamebird seasonings, or your favorite sauce.

Chapter Four
The "Wild Things" — Quail, Pheasant, Guineas, Ducks

Other Quail and Pheasants

Bobwhite Quail are a popular cage breed — they look like double-sized Coturnix, and two make a good-sized serving. Males make their famous "bob white" cry, and in the spring sport a jaunty feather curlicue on their heads. These birds lay primarily in the springtime, though some breeders are producing year-round laying. Bobwhite will set eggs, but you'll get a larger hatch if you incubate.

The birds can be kept in chicken-sized cages, one male to one or two females. Feed them game-bird foods, or chick starter, plus plenty of greens and chopped vegetables.

Butcher like chickens, but make the cut up the back as for Coturnix so that the presentation on the plate is attractive.

Pheasants are even more like chickens in their requirements than quail are — same cages, same feeders and waterers, but they do need game-bird feed or chick starter. Like quail, pheasants are seasonal layers, producing a couple dozen eggs before "going broody." One male will suffice for a handful of females if they are kept together. These birds will make attempts to fly and can damage their heads against cage tops — trimming one wing's feathers will discourage this.

Butcher these birds as you would a chicken. Serve pheasants "under glass" for that perfect elegant touch!

Guineas

These birds, with their polka-dot plumage and the dinosaur-like boney knob on their heads, seem primitive and wild. They also have an unusual call: males say, "buckwheat!," and females just squawk. Some homestead magazines have declared guineas "watchdog birds" because they call excitedly

Backyard Meat Production

when strangers arrive — they also screech whenever a bird flies overhead, or a nearby dog barks, or you walk into their area with feed. But they will let you know when anything out of the ordinary is going on!

Guinea chicks are called "keets," and need the same care that baby chickens require. They should be fed chick starter or game-bird feed, and have daily access to fresh water and greens. After the keets feather out, they can be transferred to regular chicken cages or pens. Guineas are excellent fliers, so pens should have wire tops to prevent escapes. These birds also prefer to roost in trees at night, so make their roosting area up off the ground.

Guineas lay a medium-sized brown egg, which is almost indistinguishable from a chicken egg, during the spring months, totaling about 30-40 eggs for a season. Some guinea hens will set, but more certain hatches can be guaranteed by using an incubator. Spring-hatched birds will lay the following year.

Butcher guineas, pheasants and larger quail just as you would a chicken. These birds all benefit by aging in the refrigerator for 48 hours before cooking or freezing.

Ducks

Ducks are very dirty birds — they really do need water deep enough to poke their heads into, and then they'll try to swim in it. Their wastes are liquid and very smelly — and most breeds of ducks quack loudly and at all hours, making them annoying and nerve-wracking. Their meat is greasy if not cooked correctly, and their eggs taste like dirt.

Chapter Four
The "Wild Things" — Quail, Pheasant, Guineas, Ducks

Having said that, let me now correct myself. That paragraph describes ducks, all right — but not all ducks, and not any ducks under certain circumstances. The typical breeds of commercial meat-type ducks include the white feathered Pekin, and a wild-looking variety called the Rouen. Both of these birds can reach a solid 10 pounds in five months or less, and produce an extremely rich meat. Raised in a backyard setting, you will find that your birds pretty quickly match that first paragraph — at least, we did when we tried to raise those breeds. However, when raised on a larger acreage where the ducks have a pond to swim in and their own pen, ducks are clean and pleasant if still a little noisy.

But now let me introduce you to the miracle duck — a bird that can get to 16 pounds at maturity, that produces a flavorful but not-too-fatty meat, and that (miracle of miracles!) doesn't quack!! This is the Muscovy duck, a weird-looking large bird with a red fleshy growth on its face and head.

The "crenellated" head of a Muscovy duck.

Backyard Meat Production

Muscovies are calm, quiet birds with feathers that lack the abundant oils of other duck breeds. Instead of a raucous quack, they make a hissing sound.

Males are significantly larger than females, and have larger knobby decor on their heads. Females lay abundantly, for ducks, giving several dozen eggs before going broody. Other breeds of ducks incubate their eggs for 28 days — Muscovies take 35-38 days of incubation before a hatch is complete... to the point where you're certain that all the eggs in the nest have gone bad. Although they enjoy a good swim, Muscovies can tolerate no more than that "head dunking" quantity of water. But, they do have liquid and smelly wastes — oh, well, I guess nobody's perfect. Just plan on spending more time cleaning out the duck area.

If you are seriously partial to duck eggs, the ideal breed for you is the Khaki Campbell, a breed raised to produce dozens upon dozens of eggs yearly — 200 eggs per female is not uncommon. These don't set, but they do produce a smallish meaty table bird.

Cage-raised ducks of any breed still need their dunking water, but they do well on any typical chicken feed as long as the feed CONTAINS NO ANTIBIOTICS. Antibiotics will do in your ducks, so be certain to read the label and reject any feed which contains them (aureomycin is an antibiotic).

The dampness and smell problem can be somewhat resolved by cleaning the duck area or cage every day. If the ducks cannot touch the ground, you can sprinkle plain agricultural lime (calcium carbonate) over their wastes to reduce the odor — but if the ducks walk on this for any length of time, it can "burn" their feet.

Chapter Four
The "Wild Things" — Quail, Pheasant, Guineas, Ducks

Purchase and Health

All of these varieties of birds tend to be very hardy and resistant to illness. The best approach, should you find a sick-appearing bird in your flock, is to remove and destroy it immediately. This prevents the spread of disease.

If you order these birds from catalogs, you can expect to find prices quite variable. Coturnix can't be mailed as day-old chicks because they are so fragile, but you can order fertile eggs for about $25 for 25. Bobwhite eggs cost about the same as Coturnix. Guineas and pheasants cost about $75 for 30 chicks.

Ducks also vary by breed, typically running about $70 for 30 ducklings. With prices this high for day-olds and for eggs, it is certainly worth your while to invest in a good incubator and keep the game birds propagating right on your property.

Don't Raise....

This brief section is a warning against raising two common types of birds in a backyard meat garden: turkeys and geese. Both produce large, oven-sized, tasty animals, which are quite popular and familiar. But both are difficult to raise well on a small property. Turkeys, in particular, are very sensitive to temperature changes and to disease — plus you have to order a minimum of twenty turkey chicks from most hatcheries (at about $3 each!). With supermarket turkeys of generic brands costing less than $10 for adult birds, you will most certainly spend more money raising a turkey than it will be worth.

The same can be said for geese, except day-old goslings run $5 to $7 each! In addition, geese are incredibly noisy and your nearby neighbors will band together to have you driven from the area if you keep these large birds.

On the other hand, if you have one-quarter acre or larger, these bigger birds can fit in very well, provided they receive extra attention as youngsters. Turkeys will need the medicated feed, and the geese shouldn't have any — otherwise, raise them as you would chickens with extra warmth and chick starter feed. The turkeys will benefit by the addition of milk to their grain; the geese will appreciate chopped greens from day one.

Turkeys and geese lay as yearlings. Turkeys of the large-breasted, white-feathered varieties cannot mate because of their awkward size, so you'll need a male of a more athletic breed (Bourbon Reds are good) for your white females. Turkeys don't set eggs.

Geese require swimming water in order to breed comfortably (a wading pool will do), and they will carefully hatch and raise their own young. Geese mate for life, so if you're going to eat them, choose unmatched "teenage" young birds for the oven.

Geese are aggressive when protecting their young, so keep your own little children away from these birds.

Butchering

The game birds from pheasant to turkey are all treated as you would a chicken (see Appendix). Water fowl need to be dunked in boiling water, and then swished up and down repeatedly for about one minute to allow the water to penetrate into the oily feathers. Then, proceed as for chicken.

Recipes

DUCK A L'ORANGE

1 or 2 fresh aged ducklings, prick skin if whole
2 quartered, cored tart apples

Orange Sauce:

½ cup orange juice
¼ cup sweet vermouth, or red wine, or lime juice
1 thick slice of yellow onion
10 juniper berries, crushed
sprigs of fresh parsley
½ teaspoon anise seeds

Mix sauce ingredients together. Pour over the ducks, which can be quartered, if you wish. Marinate in a refrigerator for six hours or overnight.

Preheat oven to 350 degrees. Roast duck for 60 minutes, but test for doneness after 30 minutes and every ten minutes thereafter. Place duck pieces or whole ducks on a rack. If whole, place cut-up apples inside. If cut, add apples during last 20 minutes of cooking.

PEASANT PHEASANT

1 pheasant or 1 guinea
1 large green cabbage, cored and with leaves separated
2 carrots, peeled and cubed
2 onions, peeled and chopped into large cubes
2 tart apples, cored but NOT peeled, cubed
¾ pound kielbasa or other sausage, cubed
2 cups red wine
2 cups beef or pork stock
salt and freshly ground black pepper

Sauté the vegetables with the sausage until they are lightly browned. Add all ingredients to a large crock pot or Dutch oven, putting a thick layer of cabbage leaves in first, then following with the legs of the bird. Follow this with bits of sausage mix, more leaves, and the rest of the bird. Cover with remaining sausage mix, and cabbage leaves.

Add the wine and the stock sauce. Cook on medium for the crock pot, and low heat if on the stove or in the oven. Keep on slow heat for two hours, checking every now and then to make sure the liquids haven't been used up. Add more wine and stock as needed.

When finished, serve pieces of the bird on leaves of cabbage. Serve oil-baked potatoes on the side, along with a fresh salad. Save leftover deboned meat and the vegetables for tomorrow's soup!

Chapter Five
Porky's Pals — Mini Pigs

Sausage. Bacon. Chops... for a confirmed meat eater, these words bring up wonderful images and fragrant aromas. There is no creature on earth better suited to being the source of these glorious gustatory wonders than the humble pig.

The barnyard bacon factories we are most familiar with, though, can easily reach 500 pounds at maturity! Typically, a six month old porker outweighs an average man at 200 pounds... way too big and strong for a backyard garden.

What's a confirmed home-garden sausage-lover to do???

The Breed

You've already heard of these pigs; perhaps a neighbor even owns one. Pot-bellied pigs are cute, friendly, clean, quiet, and easy to keep on a small property — and perfectly edible. They arrived in the US in the 1980s from Southeast Asia, where the pigs were a dietary staple.

Typically these animals reach 100-150 pounds in about a year and aren't more than 20 inches high at the shoulder. Females can breed at six to seven months, and produce between four and 12 piglets weighing a pound each at birth. Pregnancy lasts about 114 days, just under four months. Pet-breeders recommend spaying and neutering piglets at about three to four months of age, but when you're producing these animals for your table, you can neuter your young males at two weeks, and plan to butcher the girls at around four months of age. One male, a "boar," can be sire to a dozen or so female "sows" — but unless you eat nothing BUT pork, you'd be better off limiting your sow herd to a maximum of three.

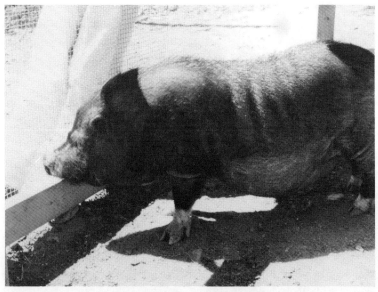

A full-sized pot-bellied pig — solid as a rock, but only twelve inches high!

Chapter Five
Porky's Pals — Mini Pigs

These animals are relatively bright, something like dogs, have an excellent sense of smell, but very poor eyesight. They root in the ground, too, so their pen should have access to some dirt. They will train themselves to use a corner of the pen as a toilet, and this should be cleaned daily just as you would a dog pen. During very hot weather, the pigs appreciate a wading pool or muddy hole to cool off in.

I would recommend avoiding pet-store piglets, and going directly to a breeder — they will probably cost less, and you can be assured that neither male nor female piglets have been neutered or spayed. All-black piglets will cost less than colorfully marked ones. If you think you might be selling pet pigs, as well as eating them, get a colorful male and black females — a percentage of your piglets will be colorful pet quality. Unless you REALLY want to raise these animals to compete in shows or influence the breed, you won't need pedigreed animals.

You may expect to spend $100-$300 each for your start-up animals, although in my area $30 per head is more common. Shop around.

A Caution...

A lot of people have forgotten where their bacon comes from. Pot bellied pigs are often treated like members of a person's household — and breeding animals may even live in the house with the human family. I would spare this pig owner an unwelcome dose of reality, and not bother to mention that I was planning to eat the offspring of their "baby."

The Pig House

A factor to keep in mind when planning a home for your porkers is that the boar should be separated from the sows in his own pen most of the time. Boars that run loose with sows can become aggressive, and will sometimes injure or kill piglets. Pregnant sows can share a community pen, but the piglets will do better if you have a separate "birthing pen." Schedule the sow breedings so that they're not all having babies at the same time, and you'll only need one maternity hospital.

A six by six foot pen is perfectly adequate for the boar. Bred sows can be community penned in an area 15 x 15 feet, though the more space they have, the happier and less destructive they are. These pigs prefer warm weather, and can tolerate quite a bit of humidity. If you have freezing weather, snow, or severe winters, the pigs will require draft-proof housing — a well-insulated dog house, with a moveable flap that covers the doorway, is perfectly acceptable. An unheated garage would also be suitable. If the weather is cold, the pigs will eat more to keep up their body temperature.

Ideally, a pig pen should have stout fencing that extends six inches under ground. You can find "pig panels," a very-heavy-gauge wire-section about 12 feet long, at livestock supply stores for about $12 each. These are suitable if they are securely attached to well-mounted posts.

Feeding Pigs

Pigs are omnivores — eaters of everything — just like us. Their rooting capacity makes them hard on lawns, shrubs, and

Chapter Five
Porky's Pals — Mini Pigs

they will wreck your potato garden in minutes if they get into it! This rooting ability can be put to work for us, though, if you want your vegetable garden soil turned over — just put the pigs into it for a couple days. They'll not only turn the soil, they'll also manure it, remove grubs and insects, and eat up leftover weed roots or things you missed harvesting. Plus, the pigs really enjoy it.

There are bagged pig feeds, designed especially for the pet pot-bellied pig. Supposedly, some pig owners taste this stuff themselves to be sure it will "taste good" to their pets. Pigs will certainly eat this and do well on it — but the cost is so far out of line for what is required to raise healthy pigs, that I can't in good conscience recommend it.

Once again, find a livestock feed store in your area. A fifty pound bag of commercial pig feed (pig starter for the little ones) costs about $4-$6 in the Midwest, perhaps twice that on either coast. This supplies all the nutrients the animals require. Add some fresh greens, fruits (even ones that have "gone off"), vegetables, your table leftovers, cooked meats, and so forth, and your pigs will prosper.

Young piglets can also do quite well if fed on cracked corn (broken kernels), with fresh milk added and soaked for an hour or so until the feed has a sloppy consistency — this is the old time "hog slop" that raised our ancestor's bacon. Make sure they also get the fresh foods, too. Don't feed a ration high in soy or soybeans — it makes "flabby pork," meat that is excessively soft.

A 150 pound boar may need two to three pounds of feed daily, in addition to greens and fresh foods — or he may get excessively fat on that amount. Vary the diet so that your breeder stays healthy but sleek. In a pinch, you can substitute cheap high-grain dog food, those made with a base of corn meal, in the pigs' diet.

Sows need about two pounds daily, increasing gradually up to two and one-half pounds during pregnancy and lactation. Babies should get all they will eat. Two or three feedings daily helps the animals use their grain more efficiently, but they can get by on a single daily meal.

Make sure that the pigs have a constant supply of fresh water, provided in a heavy crock or trough, but no more than two inches deep when piglets are around. A wading pool will get muddy quickly and shouldn't be the only drinking water source.

Breeding and Piggies

Sows come into season on average every three weeks, and remain in "heat" for about three days. This is the only time she can be successfully bred. Unbred sows get grumpy and irritable, a piggy version of PMS. There will be a slight discharge from her vulva, and a willingness to "stand" for the boar when she is receptive. Turn her in with the boar, and the task will be quickly accomplished. If the two fight viciously, or express no interest in each other, remove the sow until next heat. A single breeding is usually sufficient, but you may wish to rebreed the sow again in several hours to be certain of success. Mark this date on your calendar, then figure 114 days ahead for the due date.

About a week before the babies are due, prepare the nursery. This pen can be six by six feet, and should have a

Chapter Five
Porky's Pals — Mini Pigs

low railing nailed around the inside about four inches off the ground — this allows a safety zone where little pigs can run to and hide beneath when mama is moving about. The birthing pen should be warm, about 75 to 85 degrees (a heat lamp — $12 — suspended above the pen works fine).

Piglets are generally born without any complications. If the sow appears to be having trouble expelling one, you may *gently* lubricate the birth canal with KY-Jelly applied while wearing a fresh latex glove. When birth complications of healthy pigs take place, it generally involves a single stuck piglet. Most often, the sow pushes it out successfully — but if she's been working on getting one out for over an hour, you may need to reach your latex-gloved hand into the birth canal and *gently* dislodge the offending piglet. This is VERY RARE.

The sow will clean the piglets and they will be nursing before you realize it. Dispose of any dead, badly deformed, or mashed piglets. If you should lose the sow, or there are more piglets than she has "spigots," you can feed the piggies on baby-bottles using "sows milk replacer," available at feed stores. Make sure there is fresh dirt in the pen for the youngsters to eat — it is their first source of dietary iron. Lacking clean fresh dirt, the piglets MUST have an iron shot (see Resources for veterinary supplies) within three days of birth.

At two days of age, the piglets will be lively and energetic. Today's the day they should get their teeth trimmed. Use a pair of sharp wire cutters, and snip off the needle-sharp tips of the piglet's teeth, being careful not to snip their lips. They won't like their first dentist visit, and will squeal piteously from being handled. This is not a harmful procedure, though, and actually saves piglets from being badly chewed and mauled by their fellows.

At ten days to two weeks of age (though some breeders wait as late as three months), male piglets should be castrated. This prevents unwanted breeding, fighting, and the development of the "boar odor" which infiltrates and unpleasantly flavors the meat. Neutered males become "barrows."

Hold the piglet between your knees with its head down and rear end toward you — or have someone else do the holding. Using a veterinary scalpel, an x-acto knife, or very sharp pocket knife, make a slit in each side of the pig's scrotum. The piglet will squeal, so be prepared to keep going anyway. Withdraw the testicle from that side, and scrape or rub the cord it is attached to until it breaks. Don't cut this cord cleanly, because that will cause bleeding and more pain to the piglet. Repeat this process on the other side. Dispose of the testicles. Piglet's wounds can be treated with any good veterinary antibiotic powder or spray. The powder comes in shaker-type canisters; when applied, the fluids at the wound site form the powder into a thick paste which both covers and protects the wound. The spray — which is generically called 'blue spray' because it leaves a vivid royal blue coloring on sprayed sites — is similar in action to human-type first-aid sprays. (And, human-sprays are quite effective on piglets, too — just more expensive!) Some sprays include pain-killing medication, antibiotics to kill germs, and even an insect repellent to protect the animals from pests as they heal. Both products work equally well, but in very humid environments the spray will be more long-lasting in effect. The new barrow

Chapter Five
Porky's Pals — Mini Pigs

will run around for a few seconds, then will remember that there's food nearby, and will appear to have forgotten the whole incident. Healthy piglets will heal rapidly. If there is any excess swelling or exudation from the cuts, the pig should have a tetanus and penicillin shot (see Resources for veterinary supplies). Studies have shown that animals castrated after receiving pain-killers actually did worse and had more complications than animals that received no pain-killers.

If you wish, you can vaccinate your piggies and adults. At six weeks of age, recommended shots are: rhinitis, erysipelas, leptospirosis, and transmissible gastroenteritis (TGE). You may be able to find these combined into a single shot — check vet catalogs and order the vaccines and syringes by mail. Booster shot the same at nine weeks of age, and again at six months. Then, once a year the same shots can be given. Your breeding adults can live well into their teens.

Health Care

Pigs which start out healthy seldom suffer from any ailments. Don't buy ones with deformed snouts, lameness in the limbs, or a bulging "navel hernia," and you'll save yourself a great deal of anguish, even if these animals are cheaper.

Pigs have a very coarse but thin hair coat. Brushed every now and then, the pigs remain clean and shiny. However, pigs can get sunburn and frostbite from not having proper housing and shade — and they can pick up dandruff, mange, and lice. Excessive skin dryness and dandruff can be remedied by adding a quarter cup of cooking oil to the animal's diet — as well as by rubbing a little oil into its skin. Mange and lice can be treated by veterinary powders, rotenone powder, and diatomaceous earth (garden stores sell the last two). A good greasing with cooking oil also will kill mange, mites and lice, besides giving the piggy a pleasant rubdown.

Pigs can get head colds, influenza, pneumonia, and "atrophic rhinitis" (a runny nose), just like people. Treat with 1000 mg of vitamin C, and 50,000 units daily of vitamin A for one week ONLY, adding the tablets to the daily feed. Pneumonia may require penicillin shots (see Resources for veterinary supplies). Pigs which repeatedly come down with respiratory ailments may be exposed to conditions in their pens which foster problems — get down on your hands and knees and take a good sniff around the pen. You may find areas where "ammonia" buildup stings your eyes and takes your breath away — and this may be the source of the piggy's problems. Clean the pen!

Adult pigs that will be your breeders for years to come should also have a rabies vaccination (for your protection and theirs), and an annual tetanus shot.

All the medications I've mentioned can be ordered through veterinary supply catalogs. If you haven't given shots before, buy a needle and syringe and practice drawing up water into the needle, tapping out the air bubble, and injecting an orange. Pigs can receive shots in the shoulder or the hind leg.

Hold the limb steady. Swab the injection site with alcohol. Use a clean, sterile needle and syringe. Put the needle into the muscle, pull the plunger back to be sure there is no blood vessel involved (if you see blood, pull the needle out and try again), then inject the solution. Pull the needle out quickly and dispose of it where no one will be accidentally "stuck." All done!

If a subcutaneous, "under the skin," vaccination is called for, pinch up a bit of skin on the shoulder or neck, swab with alcohol, and inject as in the previous paragraph. Once the needle is in and the plunger has been pulled back to show no blood, you can release the pinch of skin... then inject.

Harvesting the Pork Garden

Young pigs can be dispatched in the same way rabbits are: a sharp rap on the forehead with a ball-peen hammer or a stout stick. The pig will drop to its knees or fall to the side. Immediately, run a sharp knife along the jaw line deep into the throat to slit the jugular vein. The animal will "bleed out" in a matter of minutes — residual reflex muscle motion may take place, but the animal is dead. Remove the head.

Pig skins can be cut off as a whole section, or the animal can be dunked into a boiling water tub for three minutes and then scrape the hairs off. The former method produces good meat, but you'll lose some of the fat. The latter method is the one our ancestors used, because it helped protect the meat during unrefrigerated storage.

Proceed with the butchering, as outlined in the Appendix.

Pork fat, lard, can be collected and then "rendered," or melted down into a pure white fat that makes the finest pie crusts in the world. This fat can also be made into soap using recipes found on supermarket cans of lye. If you don't wish to eat it or make soap, you can also feed it back to your pigs (after rendering), or to your chickens to boost their growth and egg production. Dogs and cats really enjoy it, too! Store rendered lard in the deep freeze or in air-tight containers in cool temperatures.

Recipes

Bacon is made from the fatty strips of meat overlying the animal's rib cage. Take the whole rib side, slide off the bones so that you have a flat thick piece of meat and fat. Supermarkets and sporting goods stores carry spices for making jerky and bacon, or you can find "cures" for meat in the spice section of markets. Apply this according to package directions. Let the bacon marinate the required time, then, if you want, you can put it into your electric or regular smoker for a day's worth of extra flavoring. When done, refrigerate until firm, and slice thinly or thickly — package and then freeze it hard.

RENDERING LARD

Take fresh fat and remove all visible meat. Cut fat into one or two-inch cubes and place on a wire rack over a baking pan. Put this in the oven at 200 degrees, so that the fat melts slowly and drips into the pan. Check after two hours. Spoon any "cracklins" or crunchy bits out of the hot oil in the pan. Pour the hot oil through a sieve or several layers of cheesecloth into a HOT, DRY, AND CLEAN mayonnaise type jar (if it's cool it will crack; if it's damp, it will ruin the fat). Cap the jar with a canning lid and band immediately while still hot, and the lid will seal. If sealed, store in a cool dark place until you need the lard — it will keep for a year if kept cool and unopened. If unsealed, wait until it is cold, then refrigerate or freeze. Will keep for three months.

TRADITIONAL ENGLISH ROAST PORK

1 or 2 pork loins, skin left on but scraped of all hair
vegetable oil
coarse salt
sprig fresh rosemary, or 1 tablespoon dry leaves
6 large tart apples
1 cup dry white wine (optional)
1 cup beef or pork stock (made from bullion cubes)
salt and freshly ground pepper
sprigs of parsley or watercress

Make sharp scoring cuts through the skin, following the grain of the meat, and rub the skin with oil and coarse salt. Put the rosemary on the rack in a roasting pan and lay the meat on top of it. Roast for two hours or so at 350 degrees.

Core the apples but don't peel. Put into the roasting pan with the pork during the last half hour of cooking time. When done, remove pork and apples to a serving plate and garnish with parsley or watercress.

Drain the fat from the pan into a sauce pan. Pour the wine into the roasting pan and heat for about five minutes; add this to the fat. Now, add stock to the fat and wine mixture and boil hard for three minutes. Strain this into a gravy boat.

Serve with boiled new potatoes and peas for a truly British feast!

Backyard Meat Production

82

Link, a bacon-bearing boar from Lake Tibbals, Washington, amply demonstrates that pot-bellied porkers can pack on plenty of poundage!

Chapter Six
The Milk and Meat Factory — Mini Goats

When the meat garden graduates up to larger space, the gardener can begin experimenting with new varieties — just like the vegetable gardener, who finds his little plot enlarged. With more ground, the vegetable gardener can grow those wildly twining plants that need to run along walls and fences. The meat gardener can graduate up to slightly bigger animals, with greater production capabilities....

And finds these miracle meat makers! Miniature goats come in two varieties: pygmy and dwarf. Both are walking meat and milk factories, which can do very nicely in small flocks on a quarter- or half-acre property, or even on larger properties!

Pygmies and Dwarves

These remarkable animals originated in West Africa. Both were imported originally for zoos and "petting farms" because of their tiny size and amazingly friendly natures. They share many characteristics in common, such as:

Maturity: Females begin breeding before they are six months old, and males are fertile and willing as young as two months! (This is even earlier than rabbits!)

Offspring: Both breeds are known for multiple births, with three and four per litter not uncommon. Babies generally weigh two pounds at birth and are up and nursing rapidly.

Breeding: Both breeds are "aseasonal" breeders, that is, they will reproduce any time of the year. Typically, a spring "kidding" or birthing will be followed by three or four months of milking — and then the goats will rebreed and produce babies later in the year, followed by more milking.

Milk: Does of either breed can give a quart of milk daily — so that three or four does will provide a full gallon of milk every day. Not only that, the milk provided by these animals is as rich and delicious as half-and-half cow's milk, with about 6% butterfat. It can be frozen, used for homemade ice cream, or even made into cheese or cottage cheese.

Meat: Young goats of these breeds are tender and flavorful, with a texture and taste resembling young beef. The pygmy goat is more heavily muscled than the dwarf, giving junior-sized "hams"; while the dwarf is considered to have more "dairy character" and potential milking capacity. The carcass is large enough to provide a meal for a hungry family, but small enough to fit in the freezer or refrigerator without taking out shelves.

Personality: These goats are very oriented toward people, friendly, and inquisitive. Pygmies are sometimes called "tree goats" because of their climbing agility. Alert and good natured, their small size makes them ideal pets for children.

Chapter Six
The Milk and Meat Factory — Mini Goats
85

Doe Pygmy goat showing her well-rounded and meaty potential.

Relative size of a Pygmy goat.
(Grand Champion Five R Rocky standing next to young Tony Moss.)
(Photo courtesy Tyny Goat Ranch, Oklahoma City, OK.)

Housing

Both breeds are hardy and can tolerate weather changes. However, they do need a shelter from wind and rain (and snow, if you get it!), and they MUST have strong, tight fences.

Shelter can be as simple as dog houses with hay on the floor, or can be a regular three-sided wind break where a group of goats can gather. Sheds and garages can be easily adapted to these breeds, giving the goats access through a doorway or opening in a wall, into a hay-bedded area. Each adult goat will feel comfortable in a three foot square area, even if a handful of babies share their space. Four adult goats with young wouldn't need more than an area ten feet by ten feet as a shelter, as long as they have an outside "play yard" to frolic in.

Does that are kidding can be separated into a private 3 x 3 foot nursery pen, so that they can properly bond and so you can make sure each baby has nursed. For a small flock, this probably won't be necessary — does will seclude themselves into a private corner during the birth.

Flooring materials need to be cleaned out in spring and fall, at least, and more often if they get very smelly. Ammonia buildup can take place quickly if the floor underneath the bedding is concrete, and ammonia causes all manner of respiratory and eye problems. If you get down on your hands and knees to "goat level," you can sniff around and make sure there isn't too much heavy odor.

If ammonia is a problem, first clean out all the flooring material (aim this toward vegetable or flower gardens, after it

Chapter Six
The Milk and Meat Factory — Mini Goats

has "rested" for ten days to two weeks). Rinse or wash the floor and allow to dry. Apply a dusting layer of agricultural lime (calcium carbonate) to the concrete and cover thickly with straw or sawdust. The lime helps counteract the ammonia buildup, but animals' bodies shouldn't come in direct contact with lime for any extended period of time.

Feeding the Milk Garden

Dairy-goat rations, a pre-mixed feed that is 16-18% protein is suitable for these goats ($6 for 50 pounds in the Midwest). Does "in milk" should receive about one half to one pound of grain daily, along with access to hay or pasture. You can also vary this grain, providing a "super stock" which is a mixed grain and pellet combination which has molasses added ($5 for 50 pounds, Midwest) — goats really enjoy this. It looks and smells like granola! Read the feed label to be sure it contains no "urea," which can be deadly to goats.

Hay can be any good quality grass or legume such as alfalfa, which is bright and clean, with plenty of tiny dried leaves and narrow stems. This costs about $3 per bale in the Midwest — and about $10 per bale on the West Coast. Goats don't like thick straw-like hay, and often won't eat it. Each adult mini-goat can consume about one to two pounds of hay daily along with their grain, so that a single "flake" will do for three or four adults. Hay should be fed so that the goats can't walk over it — perhaps in a small feeder or behind wood "keyhole" slats so that goats can get their heads into the hay, but not the rest of them!

If you keep the kids with their dams, the youngsters will get plenty to eat by nursing and by snitching feed and hay — they don't require additional foods. If you wish to use the milk AND keep the babies on their mothers, pen the kids separate from the dams for half the day — say, morning until afternoon — then milk the moms and turn them back with the kids. You get milk, and so do the youngsters. Everybody is happy!

Goats also require year-around access to trace minerals and salt. This can be found at livestock feed stores for $7 per fifty pound bag. It also comes for the same price as a "block" which weighs about 30 pounds and consists of trace minerals, salt, and molasses binder — goats love it!

Milking

To milk these goats without throwing your back out, you'll need to lift the animal up to a comfortable table-top height. If you wish, you can prepare a ramp up to a "milking stand" which allows you to sit down while milking. Put the doe's grain in front of her, and milk her while she nibbles.

The doe can be milked from either side, or even from the back. Grasp the udder firmly and massage both sides. Using thumb and first two fingers, hold the teat. "Close" the top of the teat by compressing between thumb and index finger, and then press the next finger against the teat... this will force the milk out the teat holes. Catch this fluid in a quart jar, if possible, or any container that won't be subject to falling hair or the goat's foot. Don't pull downward on the teats — this can cause damage to the udder.

Chapter Six
The Milk and Meat Factory — Mini Goats

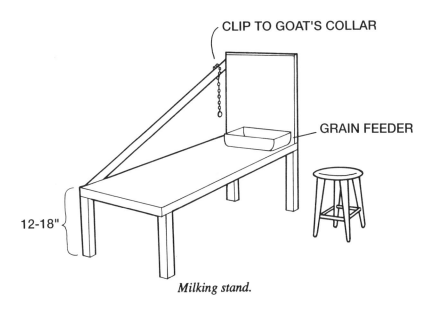

Milking stand.

Take the fresh milk into the house, and pour it through a "milk filter" or sterilized (boiled) cheese cloth into a holding jar that has been cleaned with boiling water. This filters out hair, and other junk which might have fallen into it. At this point, the still-warm milk can be used "raw," or it can be pasteurized by heating it up to 150 degrees for five minutes (don't boil it!). Or, it can be cooled in the refrigerator, and then frozen for later use.

If raw milk is refrigerated and has been handled with attention to cleanliness, it will keep for at least a week in good condition. After that, it will gradually sour and may end up separating into "curds and whey." This is perfectly normal, and if the soured milk has no off colors (such as red or green spots!), then it's perfectly safe to eat, too.

Goat Meat Gardening

Goat meat is known as "chevon." It is fairly lean, flavorful, a dark tan to reddish color when fresh, and tastes a little like "baby beef." It can be used in any recipe which is suitable for beef, venison, or even pork.

In order to produce the offspring which will grace your table, you will require at least one doe and access to a buck goat. Mini-bucks are generally very attractive animals. Pygmy bucks sport a thick ruff or "cape," which makes them look doubly stocky and solid. Bucks of either breed can have horns, which they will use in defense and for "being tough."

Bucks should have two well-descended testicles in their pendulous scrotum. A buck with a single testicle CAN breed quite successfully, but this is considered a fault nevertheless, and shouldn't be perpetuated. Legs should be straight and sturdy on both breeds, with no visible defects.

Bucks do develop some rather unpleasant habits, which are characteristic of all goats. Bucks tend to relieve themselves on their front legs and on their beards — this is normal, if a little bizarre (to us humans, anyway!). They also develop a "bucky" odor which is very unpleasant and strong — but the does seem to enjoy it. The bucky smell really clings, too, especially if you have a friendly buck who likes to cuddle up to you. Breeders have tried washing bucks off with a pine-smelling cleanser — but then you end up with bucky-pine! If you dehorn bucks as kids, you can also remove one of the scent centers on their heads... but they'll still smell, just not quite as strongly. When the does are in milk, the buck can be penned separately to prevent the transmission of this odor to the milk.

Handsome adult buck Pygmy goat.
(Photo courtesy of Tyny Goat Ranch, Oklahoma City, OK)

But you don't NEED to own a buck, as long as someone within driving distance does: bring the does when they are in heat and let the job be done at the buck's residence. It only takes a few minutes, and the cost is quite reasonable ($25-50, typically).

Does should be healthy and free from obvious defects. The udder should have two teats (nipples), and be balanced and free of lumps or hot spots. Does breed during a three day "heat," which is characterized by a slight discharge from their vulva, tail "flagging," and occasional odd-sounding bleats. Sometimes, it's hard to tell when this is taking place — but if you have a buck, he'll be able to tell!

Gestation lasts about five months or slightly less. Most often, there are no complications of birth. Some breeders report that extra-large pygmy kids have required Cesarean sections (done by a vet) to get the baby out — BUT, when you get your breeding stock, you should ask the owner about birthing or breeding problems in their line. If C-sections take place, you might consider another breeder with stock that is less likely to suffer from this.

Multiple births are common in both goat breeds, with triplets and even quadruplets not unexpected. Pregnant does at term look like they're ready to burst! Most often, the does are attentive mothers and will clean and nurse their new young before you know they've been born.

Each kid should be inspected shortly after birth. Healthy kids will wiggle and may squawk when you pick them up. Female kids have a small orifice beneath the anus. Males have an anus and a tiny scrotum. The umbilical cord may still be dangling from the kid's belly. Some breeders swab the cord with iodine solution to prevent "joint ill" caused by disease organisms found on the ground. Check the doe's udder by milking each teat separately — there is a plug in the end of unused teats. When milk flows freely from a nipple, that teat has been used by at least one of the new kids. In most cases, all the kids will have a chance to nurse. If any of the youngsters don't seem to be keeping up, you can supplement that one with a baby bottle and some "goats milk replacer" or "lambs milk replacer" (but NOT "cows milk replacer — kids can die from it). I'd keep an eight pound bag of goats milk replacer on hand ($15), just in case.

Chapter Six
The Milk and Meat Factory — Mini Goats

Kids that are weak or feel "cold in the mouth" (put your finger into their mouth), should be brought indoors and warmed under a heating pad. These kids should also receive some of the does' "first milk," a thick yellowish liquid called colostrum, which contains high levels of nutrients and antibodies to help establish the kid's intestinal balance. Milk a little colostrum out of the doe into a baby bottle (the kind made for humans is fine for these little goats, though they may need a "preemie" nipple). Make sure this is warmed to body temperature, but don't cook the colostrum. Put the nipple into the kid's mouth after the kid is fully warmed up. Freeze extra colostrum for emergencies. Tiny newborns may take two to three ounces to start with. Don't overfeed!

Pygmy goat kid, full of life and spunk.
(Photo courtesy of Tyny Goat Ranch, Oklahoma City, OK)

Kids that have been chilled and are slow to start up may also need a simple soapy-water enema. This will clear the black, tarry first stool, "meconium," out of the end of their bowel — if the meconium doesn't move out on its own accord within twelve hours of birth, the kid will die. Most healthy kids pass this on their own, but weak and chilled kids often don't.

Prepare a cup filled with warm tap water, and add a dash of dish soap, enough to make the liquid a little bubbly. A simple 5-cc syringe or an eye dropper can be used to pull up the soapy water. Put KY-Jelly or Vaseline on the end of the syringe, and insert carefully into the kid's rectum (the circular opening immediately under the tail). The kid will bleat. Instill the solution into the colon, and remove the syringe. A runny black stool will follow the solution out. Add more solution until a thick, ropy, final meconium shows up — then, the enema is done.

Set the kid down, make sure it is nursing, and keep it warm. Once the baby is up and lively, it can be returned to its mother. Watch to be sure the doe will accept the youngster.

At three to 14 days of age, young bucks that will be used for dinner should be neutered. There are several devices that can do this job, or it can be done as for young piglets (see Chapter Five). We've used elastic bands for years and found they work just fine. The bands look like a very thick dime-sized rubber band (100 cost about $3). This band is slipped on the end of an "elastrator" stretcher ($12), which looks a little like a pair of pliers. When squeezed, this stretches the band large enough to slip it over the goat's scrotum, and fits the band snugly against the animal's underbody. Feel for the

Chapter Six
The Milk and Meat Factory — Mini Goats

presence of the testicles inside the scrotum before fully applying the band — they will feel like a lump inside the skin. Then, release the handle of the elastrator and work the band off the stretchers. The band is now firmly attached around the scrotum right next to the animal's underbody. The goat won't like this, and may bleat or lay down uncomfortably. After a few minutes, he will be up again as if he has completely forgotten. Within several days, the animal's testicles will shrink from lack of blood flow, and after two weeks or so, the scrotum will be dry and will fall off. The goat has then been successfully neutered.

Neutered goats don't develop large horns, and never take on that bucky odor. They are incapable of reproducing. They will grow as well as unneutered animals.

Health Care

Beginning with well-fed, healthy breeding stock will go a long way toward preventing health problems down the line. Adult bucks and does, registered, may cost $100-$300 each. Unless you think you'd like to show the animals — or plan to sell offspring — registered animals aren't necessary or even desirable. Unregistered can be just as good eating — and may cost half as much to start with.

Ask the breeder about vaccinations. Goats that are kept in a "closed herd," that is, one that has no new introductions from other flocks and that doesn't go out to shows, may not be exposed to many possible diseases. The less exposure, the less the animals need vaccination.

However, for your adults who will produce the meat, it is wise to vaccinate with C&D types enterotoxemia and tetanus — these come together in a single shot. Typically, you'd purchase a 25 milliliter bottle of C&D/T vaccine, then draw out two cc's per vaccination per animal. The C&D/T bottle costs about $6, so you could vaccinate 12 goats for about 50 cents each — pretty cheap protection. Enterotoxemia is a diarrheal disease caused by a buildup of the products of a certain intestinal bacteria... which acts as a toxin to the goat. Usually, enterotoxemia comes on rapidly in a well-fed and healthy animal, and can kill them in six hours. There is no cure once symptoms are underway.

In regions where rabies are a problem, or where raccoons patrol your neighborhood, rabies shots should also be given to your adult breeders annually. These can be purchased through veterinary supply catalogs, too (ten shots for $15) — although in some states your home-given vaccine won't be considered "valid" if the goat is exposed to rabies. But it will still provide protection anyway.

Goats can be dusted with "Co-Ral" brand delicing powder during spring and summer, or at any time the little bloodsuckers get out of hand. This powder also works against ticks.

Our caprine friends also need to be dewormed at least twice a year — in fall after a hard frost, and in spring just as the grass is coming up good and strong. There are "bolus" or pill-type wormers which are suitable for both goats and sheep. This can be crumbled to a powder and mixed in with their grain feeding. Rotate dewormers so that you are using a different chemical base with each deworming — that way the

Chapter Six
The Milk and Meat Factory — Mini Goats

parasites don't develop a resistance to the product. "Natural" deworming remedies may also be successful if your property is not overstocked and the animals have continual access to clean food. Usually, natural dewormers contain herbs, garlic, carrot, diatomaceous earth, and other harmless substances which suppress worm activity. If you use natural products, plan to deworm AT LEAST once per month to keep the parasites under control.

If the goats have a concrete area or rocky field, they may not need to have their toenails trimmed. Should you see their hooves growing lopsided or with "pixie toes," they need a trimming. Clip with common garden shears, or even a heavy pair of scissors — make their feet look like kids' feet.

Perhaps the biggest health threat to your meat garden will be the neighborhood dogs. For some reason nobody fully understands, dogs love to chase small goats — maybe its the challenge of catching them. Big dogs, little dogs, even your own dog, may find this pastime irresistible. Neighbors can be asked in charitable terms to keep their dogs leashed or penned *please* — it's illegal to let dogs roam within most city limits. If the dogs keep showing up, local Animal Regulation authorities can be contacted to find out your options. Way out in the country, marauding dogs get the 3-S treatment (shoot, shovel, and shut-up), but city manners are a little different.

If it's your own dog who is getting into trouble, he can get lessons in obedience while you are with him... keep the dog on a leash when you go to the goats, and reprimand him if he tries to yank or run after the goats. Be sincere and serious — the dog will understand. Dogs that are raised in the company of goats generally have a different attitude toward them and don't make such a game of chasing. If the dog simply will not obey you after repeated tries, consider replacing the dog.

Goats which have been mauled by dogs are a pretty pathetic sight. As inexperienced predators, dogs don't know to "go for the throat" — they usually tear at the goat's flanks, udder, scrotum, tail, or on the ears and back. There may be many skin tears, patches of skin where the hair has been pulled out, signs that the goat blundered in its fear into fences. Eyes may be injured, legs broken or twisted from panicked dashes — and goats may actually die from fright or exhaustion, without ever receiving a single bite. Evidence of suspected dog attacks should be photographed or videotaped. If you can get a picture of the dog or dogs responsible, you're a long way toward solving the problem — since most dog owners don't want to believe that Spot or Princess could ever do such a thing.

Assess the damage. Some animals may be dead or dying; some may be alive, breathing hard, eyes wide and staring or closed, or show other signs of shock. Animals in shock may or may not recover, depending upon the severity of their injuries — call your local veterinarian immediately for his advice. Animals that are up, limping, shaky, or just plain shook up, will probably come through it physically all right... but they will be spooked at the sight of dogs in the future. Veterinary bills should be presented to the dog's owner, along with proof (pictures!) of the dog's culpability.

Chapter Six
The Milk and Meat Factory — Mini Goats

Harvest to Table

The night before you plan to butcher, withhold all feed from the animals involved, but allow them all the water they will drink. This makes the butchering process cleaner when the gut is relatively empty.

Take the goat into an area away from the sight and hearing of the other goats. Put a pan of feed down in front of the animal, and then use a heavy stout stick (an oak baseball bat works well), or a hammer to stun the goat with a sharp, hard, clunk on the back of the head. Don't try to stun the animal by hitting on the front of the head — they've got tough skulls there for butting horns.

If you've hit hard enough, the goat will fall to its knees or on its side. Quickly, use a very sharp knife to sever the jugular vein by cutting deeply into the throat along the line of the jaw — or, you can simply remove the head at this point. The goat may give the usual reflex running movements, but the animal is already dead.

The rest of the process is the same as outlined in the Appendix. Save the skin and tan it, using the recipe in the next chapter.

Cooking With Meat and Milk

Use goat meat in any recipe for beef, veal, lamb, or pork that you regularly enjoy. Then, try these for variety:

THE PARSON'S VENISON

2 rear legs of goat, deboned and opened flat
bacon drippings (fat left after cooking bacon)
1 small finely chopped onion
1 cup chopped mushrooms
½ cup chopped cooked ham or sausage
2 tablespoons freshly snipped chives
salt and pepper to taste

Marinade:
1 cup red wine
6 tablespoons port wine
3 crushed juniper berries
¼ teaspoon allspice, ground
2 tablespoons vegetable oil
3 tablespoons red wine vinegar (salad vinegar)
1 bay leaf

Sauté the onions and mushrooms in the bacon fat until the onions are soft, then stir in the ham or sausage, chives and seasonings. Let rest.

Pepper both sides of the goat legs, then spread the sautéed mixture on each and roll the legs around the mix. Secure by tying with cotton butcher's twine, or with metal skewers. Place both legs in a heavy dish. Mix the marinade ingredients and pour over the legs. Cover and refrigerate for at least 24 hours. Drain.

Use more bacon drippings, heated to a sizzle. Brown the legs on all sides. Pour the marinade into this pan, and bring the whole mix to a boil. Place all in a heavy Dutch oven or

Chapter Six
The Milk and Meat Factory — Mini Goats

roasting pan, and cook at 350 degrees for about two hours until the meat is tender. Baste from time to time with the marinade.

Serve with fresh boiled potatoes, rolls, and a salad. This tastes surprisingly like venison — enough to please the parson, I guess!

FRESH "SPREADING" CHEESE

The perfect use for that extra milk; double or triple the recipe as needed.

1 quart whole goat's milk
¼ cup vinegar or lemon juice

Bring the milk almost to a boil — there will be tiny bubbles around the edge of the pan. Turn off the heat. Immediately add the vinegar or lemon juice. Stir once or twice. The milk will separate into curds and whey. Let this sit and cool off until you can handle it with ease.

Pour the curds and whey through a boiled cheese cloth resting in a colander so that you catch the curds as the whey drains into a collecting dish. The whey can be consumed if you wish, or fed to chickens, pigs, cats, or dogs.

Take the curds in the cheese cloth, twist slightly to drain, and then put the curds into a bowl. Add ¼ teaspoon of salt per half pound of fresh cheese — or other spices and flavorings as you like (chives, garlic, pepper, sugar, strawberries, etc.).

This fresh cheese can be used exactly as it is as a spread over toast. Or, you can treat it like cottage cheese, ricotta cheese, or use it in any recipe that calls for cream cheese!

Doe Pygmy goat and her two youngsters.
(Photo courtesy Tyny Goat Ranch, Oklahoma City, OK.)

Chapter Seven
Leather, Feathers, Bones, and Fibers

Beside the delicious products of your meat garden, there are additional "side dishes" that make growing meat fun and maybe even profitable! These other attributes of meat animals add to their usefulness, especially when we make use of the leather, feathers, bones and other fibers the animals produce.

Plans and Recipes

One of the most useful and attractive items that goats and rabbits produce in addition to meat and milk is their pelt. The hide of a healthy animal is shiny, luxuriously warm, beautiful to look at, and comfortable to use. Small pelts can be stitched together to make a super-warm quilt, a one-of-a-kind vest or jacket, a wind-proof winter hat, or even a throw rug. Try putting goatskins on your carseats, for real "driving comfort"!

Here's a low-cost and low-labor method for turning those hides into useful items — I've used this on goat skins, sheep skins, rabbit hides, and even deer skins, but it can be used on any mammal hide when you want the hair to remain on the leather. An average goat pelt takes about an hour to process and a couple of days to dry; it costs only a couple dollars per skin (really) if you're doing several at a time; and it results in a soft, workable hide that can be used as-is or cut up for sewing projects.

Handling the Skins: The quality of hides you begin with will make a big difference in how your pelts tan out. Fresh hides, right off the animal, should be cooled immediately. Trim off any flesh and scrape visible fat from the hide. You don't have to scrape it completely, though — that will take place later, and be much easier at that point. Place the skin in the shade, lying completely flat with the fur side down, preferably on a cold concrete or rock surface.

When the skin feels cool to the touch (a half-hour or so), immediately cover the fleshy side completely and thoroughly with plain, non-iodized salt. Use three to five pounds per two goat skins. Three pounds of salt costs less than a dollar, and this is the first treatment for a hide which makes the next steps possible. If skins aren't salted within a few hours of removal from the animal, you might as well forget it — they will have already begun the process of decomposition and will probably lose their hair during processing.

At this point, because you're probably busy dealing with the animal carcass and don't need to worry about the pelt, you should leave the skin in a protected spot to dry. You may tack it lightly to a tree, fur side down; or to the side of a garage, or lay it flat wherever is handy. Add salt again if you've lost a lot in moving it; the salt will draw moisture from the skin and liquid may pool in low spots. Just add more

Chapter Seven
Leather, Feathers, Bones and Fibers

salt. Transport the skin flat — you can stack several, if you keep a good layer of salt between them.

We've had trouble with neighborhood cats and other predators gnawing the edges of skins that were too easy to get at. Put the hide so it is out of reach, even if that means laying it in the bottom of a cage and locking the troublemakers out! You don't need to stretch the skin yet, just make sure that it is perfectly flat, with no curled up edges. Let the skin dry until it is crispy. This may take a few days to a couple weeks. When completely dry, the skin is very "stable" and won't change or deteriorate appreciably. We've found salted goat skins which we'd forgotten in a barn for a couple years — they tanned up as nicely as if they were brand new.

If you're able to tan within a half-hour of taking the hide off the animal, you may skip the "salting" step.

Equipment: You'll need two large plastic trash cans, about 30 gallon size, and one lid. In addition, have on hand measuring cups, a wooden stirring stick about four feet long; staple gun and staples OR hammer and small nails; wire bristle brush; and a wood rack (or stretcher) to tack the animal pelt for drying.

For a sufficient quantity of tanning solution to tan four to six goat skins or 10 rabbit skins, use these ingredients (cut the recipe in half for fewer skins):

7 gallons of water
16 cups plain salt (not iodized)
2 pounds (16 cups) bran flakes
3½ cups battery acid (from auto parts store)
one box baking soda
can Neat's Foot oil

Mixing the Solution: A couple hours before you plan to tan, take three gallons of water and bring to a boil. Pour this over the two pounds of bran flakes. Let this sit for an hour. Strain the bran out, saving the brownish water solution. Discard the bran flakes (or save to feed to critters).

Next, bring the remaining 4 gallons of water to a boil. Put the 16 cups of pickling salt in your plastic trash can. Pour the water over the salt, and use the stirring stick to mix until the salt dissolves. Add to this the brown bran liquid. Stir.

When this solution is lukewarm (neither hot nor cold, comfortable to the touch), you are ready to add the battery acid. Keep the box of baking soda right next to you with the top open. Very, very carefully pour the battery acid along the side of the trash can into the solution — don't let it splash, if you can help it.

If you get battery acid on your skin, use plenty of cool running water to rinse it; then apply the baking soda; rinse again. If battery acid splashes into your eyes or mouth, immediately run cool tap water on the spot for AT LEAST ten minutes. You should see a doctor right away. If battery acid splashes on your clothes, it will eat a hole into the fabric — flush with water and cover with baking soda.

Stir the battery acid in thoroughly. Keep the trash can lid tightly on this whenever you move away from it, even for a couple of seconds. It's a good safety habit.

If you have dried skins, soak them in clear fresh water until flexible. At this point, you can peel off the dried inner skin from the hide fairly readily. If you have fresh skins, use as-is. Add the skins to the solution. Stir the hide or hides, pressing down carefully under the liquid until fully saturated. Leave

Chapter Seven
Leather, Feathers, Bones and Fibers

them to soak for 40 minutes, stirring from time to time to make sure all parts of the hide are exposed to the solution. After 40 minutes, the soaking tan is complete.

During the soak, fill your other trash can with lukewarm clear water. When the time is up, use the stirring stick and move the skins one-by-one into your other trash can. This is the rinsing process, which removes excess salt from the skins. Stir and slosh the skins for about five minutes, changing the water if it gets very dirty-looking.

At this point, some people add the box of baking soda to the rinse water — there are pluses and minuses on this decision. Adding baking soda will neutralize some of the acid in the skin. This is good because there will be less possibility of residual acid in the fur to affect sensitive people. However, this also may cause the preserving effect of the acid to be neutralized, as well. You need to make the choice to use baking soda based on your own end use for the skins. If they will have contact with human skin over an extended period of time, I'd use the baking soda. If the pelt will be used as a rug, seat cover or wall hanging, I probably wouldn't.

Remove the hides from the rinse water. They will be very heavy. Let them hang over a board or the back of a chair or other firm surface to drain. If the pelt has tears or holes, you can mend them at this point. Use waxed thread (run a wax candle along the thread after it is in your needle) and a whip stitch (the kind you see on moccasins) to make a secure and relatively invisible mend. If the hole is large, you can cut a piece of the leg ends and sew it in place.

Now, using a sponge, rag, paper towels, or a paint brush, swab the still damp skin-side of the hide with about an ounce of Neat's Foot oil. It should be absorbed fairly quickly, leaving only a little oily residue. That's okay.

Drying: Tack the hide up, skin side down, to your "stretcher"... we use salvaged wood pallets. Gently pull the hide as you tack it so that there is some tension in the skin — no need to exert excess pressure or over-stretch. Try to put your tacks around the edges so that the marks won't show later.

Set the hide in a shady place to dry. A slow, steady drying with plenty of fresh air circulation and NO DIRECT SUNLIGHT produces the best results.

Cleanup: Your tanning solution can be neutralized for disposal by adding a couple boxes of baking soda to it. It will froth and bubble vigorously and release a potentially toxic or irritating gas, so give it plenty of ventilation and get away from the bucket while this is happening. When all the bubbling has stopped, about five minutes later, you can pour the mix out. Your town may have ordinances preventing you from pouring it down the drain — and all that salt would be hard on metal pipes and septic systems. I've heard of people putting this salty mix on driveways and paths where they wanted to stop weeds from growing.

Final steps: Check the hide every day. When the skin-side feels dry to the touch in the center, but still flexible and somewhat soft (before the skin is crispy-dry), take it down from the rack. Lay the hide fur-side down and go over the skin with a wire-bristle brush. This softens the skin and lightens the color. Don't brush heavily or excessively in one spot — just enough to whiten it and give it a suede-like appearance. After this, set the skin where it can fully dry, a day or so

Chapter Seven
Leather, Feathers, Bones and Fibers

longer. When fully dry, the skin is finished. Lay it on a chair, fit it onto your carseat. Doesn't that look and feel great?

Care: Skins processed by this method can't be washed without some loss of quality. Washed skins get very crisp and uneven when dried. However, you can thoroughly brush the fur side, shake out the skin, vacuum it, or wire-brush the skin side to touch up any very dirty spots. Depending on how well you rinsed the hide after tanning, there may be some residual salt in the fur — this can be drying to human skin, and may leave little white flecks around, and might even absorb moisture from the air if your weather is humid. Other than being uncomfortable or making a mess, this shouldn't be a problem — the salt eventually works out of the hide.

A Gentle Caution: Once your friends know you can tan hides, be prepared for them to bring around THEIR hunting trophies and livestock hides for treatment. If you decide to do this, take my advice: don't do it for free. Commercial tanners get $25-$45 to tan a single hide, and you should price your work accordingly — even if your return is just a case of beer. Otherwise, you'll find yourself swamped with every little skin in your region, and left with no time for anything else.

In exchange, of course, your friends can expect to get a professional, top quality job, with an up-front understanding about what might go wrong (skin loses fur, for example), and what kind of compensation you get. People are very sensitive about "that special skin." This kind of precaution will prevent any potential misunderstandings and help you KEEP your friends.

Bones and Horns

I have a bud vase that I got during the summer of 1964 in Lapland, Finland. It consists of the femur (long legbone) of a reindeer, set into a tripod base made of reindeer antler. It has withstood years and years of use, been knocked off of tables and countertops without breaking, and has that rugged natural look of something made by real human hands. The reindeer that contributed the bones also probably fed his owners that same year, as well as giving them a thick warm skin to wrap up in during those long Arctic nights.

Bones can be crafted into any number of useful and unique objects — their uses are limited only by your imagination. Buttons are particularly easy to make, and really have a "natural" look. Simply take a long bone, clean it thoroughly as follows shortly, and saw button-sized cross-section pieces from it. These can be varnished or left plain.

Ancient people made sewing needles from bones, as well as fishhooks, daggers, arrowheads, and hair combs. Today, bones can be used for artistic expression, too, in a way that our practical ancestors never imagined. Just look at some of those fancy knife handles in the specialty shops. And the cane-making business, where horns are used for the handles, has never been as brisk as it is right now.

And, let's not forget "bone soup." Long bones contain an incredibly nutritious substance called "marrow," which is the source of blood cells in the body. Fresh marrow is pink and spongy. At the ends of bone-joints, there is a transparent white plastic-looking substance known as "cartilage." There is considerable archeological evidence that our meat-loving

Chapter Seven
Leather, Feathers, Bones and Fibers

ancestors cracked bones of their prey and removed marrow and cartilage for their meals. It must have seemed pretty important to them, for them to have gone to all that trouble!

Take a set of bones from slaughtering, leave a little meat and all the cartilage on them, put them in a pot with a chopped onion, some sliced carrots, chopped celery, and a bay leaf, and simmer for six to eight hours. Strain all the solids away while the soup is still warm. This soup has flavor, nutrition, and ability to thicken when cool. The cartilage from this is the same as the stuff health-food stores tout as "shark cartilage" in its ability to provide micronutrients and other substances your body needs to preserve your own bones, and boost your immune system!

When you have boiled bones (even those left after straining this soup), you're well on the way to having "cleaned" them. Wash and rinse thoroughly after boiling. When cooled, the bones you intend to use for projects should be soaked overnight in a solution of one part bleach, four parts water. This brightens and hardens the bones, as well as removing any odors or lingering bacteria.

Horns can be shaped to any form you wish, simply by boiling the horn for about one to four hours — it will then be soft and pliable until it begins to cool. If you bend (or straighten) a horn after boiling, you must clamp or hold it in the new position until it is cool — otherwise, it will simply bend back into its original shape. You'll need heavy gloves, to hold the hot horn with, and a form or clamp to hold it while it cools. Now, let your imagination be your guide!

Feathers and Fibers

Birds of all kinds produce plumage that can be as colorful and beautiful as any art you'd find in a museum. Long feathers of various types can be displayed like flowers in a vase; soft small feathers can be attached to a stick and converted into a feather duster; fishing flies can be made from speckled and colored feathers — and there is quite a lucrative business involved in this enterprise; and let's not forget duck-down pillows, comforters, and vests — they are warm, lightweight and the down is free every time you prepare a duck for the table!

Feathers should be washed by laying them on top of lukewarm water, letting them sink when they get wet. Drain the water after it begins to cool. If you have downy ones, put the wet feathers in a pillowcase, fold over the top, and hang outdoors to air dry. Or if you have larger single feathers place them in a sheltered area where the wind won't catch them (and blow them away!) — but be sure not to hold any part except the shaft... otherwise you can mess up the softer portions.

Fibers — hair — from goats and rabbits can be used for any project in which you wish to stuff something and add warmth, such as a quilted jacket or a comforter. If you are exceptionally frugal, you can save some of the fur that rabbits use to line their nest boxes (from before babies are born, that is), and use this for spinning into yarn. You can also comb the undercoat out of your goats' coats during the springtime, when they are shedding, and use that light and soft fiber for spinning.

Chapter Seven
Leather, Feathers, Bones and Fibers
113

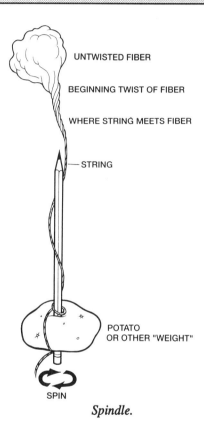

Spindle.

Spinning can be done with a simple tool, or can be an expensive hobby. At its very most basic, you'd only need a sharp pencil and a small potato to make the "spindle" (see illustration). Tie a string onto this spindle as shown. Take a small handful of rabbit hair, lay the end of the string into this, and give the spindle a turn. The rotation of the spindle passes up the string, and turns the string in the hair. The hair begins to twist, "picks up" the turn, and rolls until you have made a thin "thread." When the spindle stops, don't let it turn backwards — or you'll unwind your thread! Just continue spinning in the same direction, adding more hair as needed.

When you have a long strand of fiber, carefully roll it around the base of your spindle, beneath the "potato." This adds more weight and helps the spin become more uniform. Pile up the yarn there until you can't make any more fit, then carefully unroll the yarn and wrap it loosely around a book or board. When it is all wrapped, tie string at several places to hold the yarn in a bunch, then slip it off the book or board. Place this "skein" into lukewarm water and let it sink. Rinse if it needs it, but don't twist the skein to remove water. Hang to dry out of the sun — and you can attach a small weight to the bottom loop of the skein to stretch it a little if you like. When fully dry, the new yarn will hold its "twist" and be ready to use for any knitting or crocheting project you can think of.

It's time consuming, and you'll start out with a pretty lumpy and weird-looking piece of "yarn." But if you keep at it, you can actually spin enough yarn to make a sweater of the softest and warmest material known to man. Or, with less yarn, there're still mittens. Or socks. Or even a scarf!

Final Thoughts

There are people in the world today who have forgotten the complex relationship mankind has with our animals. These people seem to think that humans simply prey on other creatures — but, even a quick reading of this book will show that isn't true. Yes, we do use the meat of other animals, but we provide them with comfort, food, security, and affection in return.

Chapter Seven
Leather, Feathers, Bones and Fibers

There are lessons to be learned in the process of keeping animals, ones which go far beyond the simple care of critters. You'll find yourself waking up in the middle of the night when kids are due, as nervous as the mother-to-be about the outcome. You'll watch chicks hatching from eggs with awe and wonder, knowing that these new birds wouldn't exist if you didn't bring their mothers and fathers together. You will experience the cycles of nature — birth, growth, mating, death, and new birth — just as they have been happening from the beginning of time. And, you will play your part in this miracle.

Enjoy your work. Teach your children the truth about meat, by giving your animals the care they need — and by having the best barbecues in your neighborhood. It can't get much better than this!

Appendix
Butchering, Generic Style

An amazing thing about animals is that they're all basically the same underneath. All the critters in this book have livers, kidneys, similar cardio-vascular systems, lungs, brains, stomachs and digestive tracts. The birds have an additional little attachment called a gizzard (located in the neck where it joins the body) which does the work of teeth.

So, basically, butchering is the same work, no matter how big or small the carcass you're working upon. For this reason, I've written rather general instructions on butchering that could apply to almost any edible critter.

Let me suggest that you not overly concern yourself with producing "supermarket" cuts of meat — although you certainly can — because those specialized cuts are prepared using highspeed electric meatsaws on partially frozen pieces of meat. Home-raised meat can be any size, shape, or thickness — and still cook up into a mouth-watering treat.

Head: Earlier chapters on individual animals have instructed you how to stun and bleed the animal or bird; and to remove the animal's head. There are two main reasons for this: (1) once the head is off, it's emotionally easier to do the work; and (2) you don't usually eat the head (though some people do boil down pig heads for "head cheese").

Hanging: Tie loops of twine or rope around the animal's hind legs at the ankles, and suspend the animal from a strong support at a height that is comfortable for you to work at. The animal's neck and shoulders will hang down, and will continue to drip fluids.

Skinning: Carefully lift the skin away from the muscle at the back of the animal's thigh, and insert a sharp knife beneath the skin. Cut the skin ONLY(!) in the pattern shown in the diagram, lifting with one hand so that you don't puncture the muscle or gut. For very young rabbits, you may only have to separate the skin from the rear legs using your knife, then peel the hide down inside-out like pulling a glove off of the carcass. For male animals, cut to one side of the scrotum and penis.

For larger animals, begin "folding" the skin backwards from the cut line, using a DULL knife this time to peel and separate the skin. It's okay if you accidentally cut a hunk out of the muscle or poke a hole in the skin, but the fewer cuts like this there are, the more attractive the results will be. For goats, some owners actually just cut a small slit near the tail and insert a garden hose — turn on the water and the pressure separates the skin from the carcass without too much effort, as well as cooling the meat quickly. Continue until you have completely removed the skin. You may wish to remove the front feet at this time, rather than trying to skin them out. Garden clippers work on rabbits. For goats and pigs, fold the foot forward and cut the tendons, then bend the foot back-

Appendix
Butchering, Generic Style
119

wards and cut the inside tendons — then, you can twist the foot off. Cut the tail off flush with the body.

Lay the skin down, fleshy side up, and coat with a layer of salt if you plan to tan it.

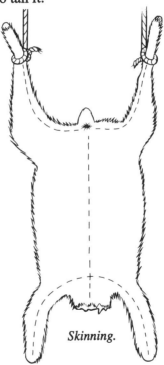

Skinning.

Gutting: For sheep and goats: with your sharp knife, cut a circle around the anus, and pull the rectum out slightly. Tie a piece of twine around it and close up the anus so no bowel contents spill onto your meat. For chickens and rabbits, this is unnecessary.

Now, again using the sharp knife, cut directly down the center of the animal's belly with the knife blade's sharp edge facing OUTWARD, not inward. *Keep one hand immediately behind the knife inside the animal's gut, to prevent accidentally cutting into the bowels.* If you do, the meat is still salvageable (wash cavity after degutted with a solution of one part bleach to four parts water), but your life will be easier if you avoid cutting in the first place.

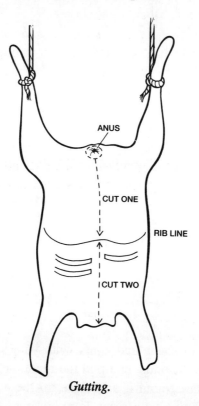

Gutting.

Cut down to the rib line — you'll hit bone. At this point, the bowels, stomach and reproductive organs (if any) will try to fall out of the open gut. Make sure you have a collecting bucket to drop these into. The rectum should slip out as a

Appendix
Butchering, Generic Style
121

whole piece with the twine still on it. If not, make a few more cuts around it until it does. The lower bowels and stomach may now rip loose from the rest of the chest contents, or there may be some bits still connecting them.

For birds, just reach in now, and pull all the remaining guts out — no need to cut down the chest.

For other animals, cut down the center line of the chest. You may need a heavier knife to cut through the cartilage where the rib bones join in the center of the chest. This is the animal's thoracic cavity, in which you'll find the liver, kidneys, lungs, heart and throat. Go ahead and pull all these out, and place in your gut bucket. If you wish, save the kidneys (which are shaped like giant kidney beans), liver, and heart. The liver has a small greenish gallbladder attached to it. Cut this out by removing liver from around it and being exceptionally careful not to cut the gallbladder. The greenish bile is very bitter and will make anything it touches inedible.

Rinse or wash off the carcass now, which should be clean of all guts, organs, and windpipe. What you've got here, is a nice muscular piece of meat. Trim any bruised-looking sections, and cut off jagged sections of neck.

Cutting: With smaller animals such a rabbits or very young goats, you can section the carcass into portion-sized pieces. It's easier to section a carcass if you cut at joints. Cut the rear legs off at the ball-and-socket joint where they meet the hip. Cut the hips off where the first vertebrae (backbone) meets it. Cut the mid-back (loin) off where it meets the rib cage. Cut the ribcage down the middle so you have a right and left "breast" section. You can also separate the foreleg at the shoulder at this point if you wish.

Larger animals can be cut into the same pieces, then cut into smaller serving sizes if you wish. In addition, you can debone meat (completely remove the bone from a piece), roll the meat up, and secure with butcher's twine, so that you end up with a lovely boneless roast.

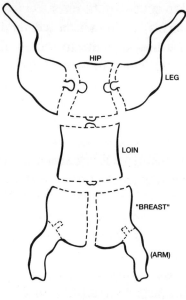

Cutting.

On your first few butchering jobs, you may end up with a bunch of little chunks and hunks of meat from various parts of the body. No problem! These are the pieces sold in supermarkets as "stew meat" and "stir fry" for exorbitant prices. Consider yourself blessed!

Remember to cut the back feet off at the ankle joint, and remove the twine.

Appendix
Butchering, Generic Style

Each section of meat should be individually washed in running cool water, patted dry with paper towels, and put in a covered container or plastic bag in the refrigerator to age. Pigs and goats can be eaten the same day, but the goat will be tough if it isn't aged. Birds and older rabbits must be aged or it will be like eating leather. Young fryer rabbits up to eight weeks old can be cooked immediately after they have cooled.

Notes and Cautions: If meat should drop and hit the ground, it is unfortunately contaminated by microbes. With a very clean area, you may be able to get away with using that meat — as long as you wash thoroughly in one part bleach to four parts water, rinse in clear water, and it is well cooked (not rare) when you eat it. If the meat falls on manure, you're risking a good case of E. Coli-caused diarrhea if you eat it. Keep your butchering area clean!

You are probably going to cut yourself at some point while preparing a carcass. Be sensible. You're working with body fluids, bowel contents, and the potential for disease or a darn good infection. If you get cut, immediately wash the cut with water, and then follow that with hydrogen peroxide (available at drug stores), which will bubble furiously and sting while it cleans the wound. Apply a good antibiotic ointment. Tape the wound securely with several Band-Aids — and wash the wound again with soap and water followed by peroxide and antibiotic and bandage after you've finished with the butchering.

You'd be smartest if you make certain your family's tetanus shots are up to date before you bring animals onto your property, and this applies especially if you get cut while butchering. If you haven't had a tetanus shot in five years, plan a trip to your doctor within the NEXT TWO DAYS — preferably the SAME DAY you are cut — and get that shot! Don't mess around with fatal incurable diseases!

Resources

Associations

American Nigerian Dwarf Organization
c/o Marsha Daniels
1433 Glocker Rd.
Sedgwick, KS 67135
www.LLNET.COM/ANDO

American Rabbit Breeders Association, Inc.
8 Westport Court
PO Box 426
Bloomington, IL 61702
(309) 664-7500
Fax (309) 664-0941
http://www.ice.net/~arba/

National Pygmy Goat Association
Terie Pleau, Business Manager
166 Blackstone St.
Mendon, MA 01756
(508) 478-5902

Chick Sources

Murray McMurray Hatchery, Inc.
Webster City, IA 50595-0458

Ridgeway Hatcheries
Box 306
LaRue 7, OH 43332

Suppliers

Caprine Supply
PO Box Y
33001 W. 83rd St.
DeSoto, KS 66018
(913) 585-1191
Fax: (913) 585-1140

Jeffers Veterinary Supply
Old Airport Rd.
West Plains, MO 65775
1-800-533-3377

Resources

Nasco Farm & Ranch
PO Box 901
Fort Atkinson, WI 53538-0901
   and
PO Box 3837
Modesto, CA 95352-3837
1-800-558-9595
Fax: (414)-565-8296
Free copy of catalog, write Dept AZ508

Sheepman Supply Co.
5449 Gov. Barbour St.
Barboursville, VA 22923
1-800-336-3005
Fax: (703) 832-2109
(These folks also carry goat supplies.)

Books and Magazines and Info

Back Home Magazine, PO Box 370, Mountain Home, NC 28758. Natural, organic orientation.

Backwoods Home Magazine, 1257 Siskiyou Blvd, #213, Ashland, OR 97520. Practical and conservative.

Cheesemaking Made Easy, by Ricki and Robert Carroll. Published by Storey Communications, Pownal, VT. 1982. Here's what to do with all that excess milk...

Backyard Meat Production

Countryside & Small Stock Journal, W11564 Hwy 64, Withee, WI 54498. This magazine is entirely reader-written and often contains useful little tidbits pertaining to small livestock.

Dairy Goat Journal, Rt. 1, Helenville, WI 53137. Aimed toward part and full time goat owners, but with excellent orientation toward the entire process of keeping milk goats.

Small Farmer's Journal, PO Box 1627, Sisters, OR 97402. A lovely oversized magazine dedicated to farming with horses and oxen, but with lots of useful material applying to smaller livestock of all types.

University of Oklahoma website: <www.ansi.okstate.edu>. This is an excellent general source for all things livestock. Look under their address but after edu add /breeds/goats (no quote marks, though) or "/breeds/pigs" for huge amounts of detail on various breeds. There's an e-mail address for each breed, as well, if you need to ask questions and get answers from real authorities!

YOU WILL ALSO WANT TO READ:

☐ **14176 HOW TO DEVELOP A LOW-COST FAMILY FOOD STORAGE SYSTEM,** *by Anita Evangelista.* If you're weary of spending large percentages of your income on your family's food needs, then you should follow this amazing book's numerous tips on food-storage techniques. Slash your food bill by over fifty percent, and increase your self-sufficiency at the same time through alternative ways of obtaining, processing and storing foodstuffs. Includes methods of freezing, canning, smoking, jerking, salting, pickling, krauting, drying, brandying and many other food preservation procedures. *1995, 5½ x 8½, 120 pp, illustrated, indexed, soft cover.* $10.00.

☐ **14187 HOW TO LIVE WITHOUT ELECTRICITY — AND LIKE IT,** *by Anita Evangelista.* There's no need to remain dependent on commercial electrical systems for your home's comforts and security. This book describes many alternative methods that can help one become more self-reliant and free from the utility companies. Learn how to light, heat and cook your home, obtain and store water, cook and refrigerate food, and fulfill many other household needs without paying the power company! Complete with numerous illustrations and photographs, as well as listings for the best mail-order sources for products depicted, this is a complete sourcebook for those who wish to both simplify and improve their lives. *1997, 5½ x 8½, 168 pp, soft cover.* $13.95.

☐ **14175 SELF-SUFFICIENCY GARDENING,** *Financial, Physical and Emotional Security from Your Own Backyard, by Martin P. Waterman.* A practical guide of organic gardening techniques that will enable anyone to grow vegetables, fruits, nuts, herbs, medicines and other useful products, thereby increasing self-sufficiency and enhancing the quality of life. Includes sections of edible landscaping; greenhouses; hydroponics and computer gardening (including the Internet); seed saving and propagation; preserving and storing crops; and much more, including fact filled appendices. *1995, 8½ x 11, 128 pp, illustrated, indexed, soft cover.* $13.95.

Loompanics Unlimited
PO Box 1197
Port Townsend, WA 98368

MGP7

Please send me the books I have checked above. I have enclosed $_____ which includes $4.95 for shipping and handling of the first $20.00 ordered. Add an additional $1 shipping for each additional $20 ordered. Washington residents include 7.9% sales tax.

Name _____

Address _____

City/State/Zip _____

VISA and MasterCard accepted. 1-800-380-2230 for credit card orders *only.*
8am to 4pm, PST, Monday through Friday.

YOU WILL ALSO WANT TO READ:

☐ **13063 SURVIVAL BARTERING**, *by Duncan Long*. People barter for different reasons — to avoid taxes, obtain a better lifestyle, or just for fun. This book foresees a time when barter is a necessity. Three forms of barter; Getting good deals; Stockpiling for future bartering; Protecting yourself from rip-offs; and much, much more. Learning how to barter could be the best insurance you can find. *1995, 5½ x 8½, 56 pp, soft cover.* $8.00.

☐ **17054 HOW TO BUY LAND CHEAP, Fifth Edition**, *by Edward Preston*. This is the bible of bargain-basement land buying. The author bought 8 lots for a total sum of $25. He shows you how to buy good land all over the country for not much more. This book has been revised, with updated addresses and new addresses added. This book will take you through the process for finding cheap land, evaluating and bidding on it, and closing the deal. Sample form letters are also included to help you get started and get results. You can buy land for less than the cost of a night out — this book shows how. *1996, 5½ x 8½, 136 pp, Illustrated, soft cover.* $14.95.

☐ **14133 THE HYDROPONIC HOT HOUSE, Low-Cost, High-Yield Greenhouse Gardening**, *by James B. DeKorne*. An illustrated guide to alternative-energy greenhouse gardening. Includes directions for building several different greenhouses, practical advice on harnessing solar energy, and many hard-earned suggestions for increasing plant yield. This is the first easy-to-use guide to home hydroponics. *1992, 5½ x 8½, 178 pp, Illustrated, Indexed, soft cover.* $16.95.

☐ **17079 TRAVEL-TRAILER HOMESTEADING UNDER $5,000**, *by Brian Kelling*. Tired of paying rent? Need privacy away from nosy neighbors? This book will show how a modest financial investment can enable you to place a travel trailer or other RV on a suitable piece of land and make the necessary improvements for a comfortable home in which to live! This book covers the cost break down, tools needed, how to select the land and travel-trailer or RV, and how to install a septic system, as well as water, power (including solar panels), heat and refrigeration systems. Introduction by Bill Kaysing. *1995, 5½ x 8½, 80 pp, Illustrated, Indexed, soft cover.* $8.00.

Loompanics Unlimited
PO Box 1197
Port Townsend, WA 98368

MGP7

Please send me the books I have checked above. I have enclosed $_____ which includes $4.95 for shipping and handling of the first $20.00 ordered. Add an additional $1 shipping for each additional $20 ordered. Washington residents include 7.9% sales tax.

Name _____

Address _____

City/State/Zip _____

VISA and MasterCard accepted. 1-800-380-2230 for credit card orders *only*.
8am to 4pm, PST, Monday through Friday.

YOU WILL ALSO WANT TO READ:

☐ **14177 COMMUNITY TECHNOLOGY,** *by Karl Hess, with an Introduction by Carol Moore.* In the 1970s, the late Karl Hess participated in a five-year social experiment in Washington, D.C.'s Adams-Morgan neighborhood. Hess and several thousand others labored to make their neighborhood as self-sufficient as possible, turning to such innovative techniques as raising fish in basements, growing crops on rooftops and in vacant lots, installing self-contained bacteriological toilets, and planning a methanol plant to convert garbage to fuel. There was a newsletter and weekly community meetings, giving Hess and others a taste of participatory government that changed their lives forever. *1995, 5½ x 8½, 120 pp, soft cover.* $9.95.

☐ **14181 EAT WELL FOR 99¢ A MEAL,** *by Bill and Ruth Kaysing.* Want more energy, more robust, vigorous health? Then you must eat food that can impart these well-being characteristics and this book will be your faithful guide. As an important bonus, you will learn how to save lots of money and learn how to enjoy three homemade meals for a cost of less than one dollar per meal. The book will show you how to shop, how to stock your pantry, where to pick fresh foods for free, how to cook your 99¢ meal, what foods you can grow yourself, how to preserve your perishables, several recipes to get you started, and much, much more. *1996, 5½ x 8½, 204 pp, illustrated, indexed, soft cover.* $14.95.

☐ **14183 The 99¢ A MEAL COOKBOOK,** *by Ruth and Bill Kaysing.* Ruth and Bill Kaysing have compiled these recipes with one basic thought in mind: people don't like over-processed foods and they can save a lot of money by taking things into their own hands. These are practical recipes because they advise the cook where to find the necessary ingredients at low cost. And every bit as important — the food that you make will taste delicious! This is a companion volume to the *Eat Well for 99¢ A Meal.* Even in these days when the price of seemingly everything is inflated beyond belief or despair, 99¢ can go a long way toward feeding a person who is willing to save money by providing the labor for processing food. *1996, 5½ x 8½, 272 pp, indexed, soft cover.* $14.95.

Loompanics Unlimited
PO Box 1197
Port Townsend, WA 98368

MGP7

Please send me the books I have checked above. I have enclosed $_____ which includes $4.95 for shipping and handling of the first $20.00 ordered. Add an additional $1 shipping for each additional $20 ordered. Washington residents include 7.9% sales tax.

Name _____

Address _____

City/State/Zip _____

**VISA and MasterCard accepted. 1-800-380-2230 for credit card orders *only*.
8am to 4pm, PST, Monday through Friday.**

"The godfather of all deviant catalogs... you won't believe your eyes. You would have doubted that books like this could even exist... This is simply the best single subversive book catalog you can get." — **Outposts**

"...Loompanics... produces and distributes some of the strangest and most controversial non-fiction titles you're ever likely to come across — books that prove truth sometimes really is stranger than fiction." — **The Winnipeg Sun**

"Their (Loompanics') catalog's self-description — 'The Best Book Catalog in the World' — is entirely too modest." — **The New Millennium Whole Earth Catalog**

"...taken in its entirety, the catalog is fascinating, offering books that provide useful information and often throw a spot light in areas never covered by big league publishers... (it) is sure to remind anyone who has forgotten what a subversive act reading can be." — **The San Juan Star**

"...hundreds and hundreds of titles you won't find at B. Dalton's, the neighborhood library branch, and definitely not at the Christian supply stores." — **The Rap Sheet**

"Loompanics Unlimited... serves as a clearinghouse for everything the government (and your mother) doesn't want you to know." — **Boston Phoenix**

THE BEST BOOK CATALOG IN THE WORLD!!!

We offer hard-to-find books on the world's most unusual subjects. Here are a few of the topics covered IN DEPTH in our exciting new catalog:

- *Hiding/Concealment of physical objects!* A complete section of the best books ever written on hiding things.
- *Fake ID/Alternate Identities!* The most comprehensive selection of books on this little-known subject ever offered for sale! You have to see it to believe it!
- *Investigative/Undercover methods and techniques!* Professional secrets known only to a few, now revealed to you to use! Actual police manuals on shadowing and surveillance!
- *And much, much more, including Locks and Lockpicking, Self-Defense, Intelligence Increase, Life Extension, Money-Making Opportunities, Human Oddities, Exotic Weapons, Sex, Drugs, Anarchism, and more!*

Our book catalog is over 200 pages, 8½ x 11, packed with more than 600 of the most controversial and unusual books ever printed! You can order every book listed! Periodic supplements keep you posted on the LATEST titles available!!! Our catalog is **$5.00**, including shipping and handling.

Our book catalog is truly THE BEST BOOK CATALOG IN THE WORLD! Order yours today. You will be very pleased, we know.

**LOOMPANICS UNLIMITED
PO BOX 1197
PORT TOWNSEND, WA 98368**

Name_____

Address_____

City/State/Zip_____

We accept Visa and MasterCard.
For credit card orders *only*, call 1-800-380-2230.
8am to 4pm, PST, Monday through Friday.